T0138939

Optical Properties of Solids

Optical Properties of Solids

An Introductory Textbook

Kitsakorn Locharoenrat

PAN STANFORD PUBLISHING

Published by

Pan Stanford Publishing Pte. Ltd.
Penthouse Level, Suntec Tower 3
8 Temasek Boulevard
Singapore 038988

Email: editorial@panstanford.com
Web: www.panstanford.com

British Library Cataloguing-in-Publication Data
A catalogue record for this book is available from the British Library.

ISBN 978-981-4669-06-1 (Hardcover)
ISBN 978-981-4669-07-8 (eBook)

Printed in the USA

Contents

Preface

In this textbook, I introduce the general point of views of the optical properties of solids and would like to give the readers an overview of the landscape of optics in solid-state materials. The book also has a special focus on optical imaging techniques. It is designed for all kind of learners, especially independent learners. It is a collection of my academic papers and researches in the past ten years and is aimed to facilitate visualization of related theoretical concepts. Problem sets have been provided with each chapter to examine the readers' understanding of each concept.

I have divided the book into two main parts. The first part comprises two sections. The first section consists of Chapters 1–6 in which I present the background of electromagnetic theory. Electromagnetic theory is based on Maxwell's equations. In this section I show how to manipulate Maxwell's equations in differential forms by utilizing vector analysis. This will also help the readers to understand the units, especially of MKS system, used in electromagnetism. They will also learn the simplest way to calculate electric field emerging from a single charge and more general ways to calculate electric field emerging from charge distributions in conductors and dielectrics under Maxwell's boundary conditions.

The second section has Chapters 7–15 in which I dive headlong into the optical theory. In this section I discuss and analyze the optical spectra from localized electronic states and go over some well-known phenomena currently under research, such as nonlinear optical response of materials. The readers will study the basic concepts of light field in order to gain insight into Maxwell's equations under vacuum or medium conditions. They will also learn the basic theorems related to the classical and the quantum mechanical treatment of optical properties of matter.

The second part is divided into two sections. The first section has Chapters 16–21 in which I give the background of optical microscopy. In this section I focus on the optical response of modern confocal microscopy on asymmetric materials. The readers will learn a measurement on nonconducting surfaces using atomic force microscopy instead of scanning tunneling microscopy on metals or semiconductors. They will learn about scanning electron microscopy to monitor atomic structure and orientation of thick or thin matter and will study transmission electron microscopy to monitor the composition and structure of ultra-thin matter.

The second section has Chapters 22–24 in which I explain about optical tomography. Tomography involves imaging the internal structure of any matter without invasion. In this section I introduce some tomographic techniques to identify the locations and profiles of matter and also concentrate on fluorescence diffuse optical tomography used as a probe in deep biological tissue because fluorescence itself offers interesting functional information of matter.

I would like to thank King Mongkut's Institute of Technology Ladkrabang, Thailand, especially the Faculty of Science, Department of Physics, for the support and cooperation provided for writing this book. I would appreciate comments and suggestions from the readers for improving this book further.

Kitsakorn Locharoenrat
Winter 2015

Chapter 1

Vector Analysis and Maxwell's Equations

1.1   Introduction

Electromagnetic theory is based on Maxwell's equations. We can manipulate Maxwell's equations by utilizing a mathematical tool called vector analysis. This tool is rather complex, but what readers have to understand here is not much. If readers understand what is explained in this chapter, they will gain insight into all the mathematics given in this book. This chapter deals with vector analysis and illustrates Maxwell's equations in differential forms.

1.2   Scalars and Vectors

A scalar is a physical quantity specified by a positive or negative number with a proper unit. On the other hand, a vector is a physical quantity specified by a direction and a magnitude. Both scalar and vector quantities can be shown by the following symbols and components (Fig. 1.1):

Table 1.1   Examples, symbols, and components of scalar and vector quantities

Quantity	Example	Symbol	Component
Scalar	Time, mass	a	A
Vector	Velocity, momentum	$\vec{A}$	(A_x, A_y, A_z)

Optical Properties of Solids: An Introductory Textbook
Kitsakorn Locharoenrat
Copyright © 2016 Pan Stanford Publishing Pte. Ltd.
ISBN 978-981-4669-06-1 (Hardcover), 978-981-4669-07-8 (eBook)
www.panstanford.com

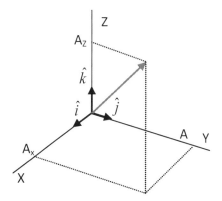

Figure 1.1 Vector $\vec{A}$ along x-, y-, and z-axes.

Vector $\vec{A}$ is given by

$$\vec{A} = A_x \,\hat{i} + A_y \,\hat{j} + A_z \,\hat{k} \tag{1.1}$$

Here, $\hat{i}, \hat{j}, \hat{k}$ are fundamental unit vectors in the x-, y-, and z-directions.

The absolute value of this vector is

$$A = \left|\vec{A}\right| = \sqrt{A_x^2 + A_y^2 + A_z^2} \tag{1.2}$$

If we define a direction in cosine terms as

$$(\cos \alpha,\, \cos \beta,\, \cos \gamma) = (l,\, m,\, n) \tag{1.3}$$

it reads

$$\begin{aligned}
A_x &= \left|\vec{A}\right| \cos\alpha \\
A_y &= \left|\vec{A}\right| \cos\beta \qquad l^2 + m^2 + n^2 = 1 \\
A_z &= \left|\vec{A}\right| \cos\gamma
\end{aligned} \tag{1.4}$$

1.3   Scalar and Vector Products

A scalar product is the length of vector $\vec{A}$ multiplied by the length of the projection of vector $\vec{B}$ onto vector $\vec{A}$, as displayed in Fig. 1.2.

From Fig. 1.2, we get

$$\vec{A} \cdot \vec{B} = A_x B_x + A_y B_y + A_z B_z \tag{1.5}$$

$$\vec{A} \cdot \vec{B} = \left|\vec{A}\right|\left|\vec{B}\right| \cos\theta \tag{1.6}$$

where θ is the angle between the vectors $\vec{A}$ and $\vec{B}$.

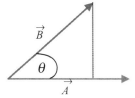

Figure 1.2   Scalar products.

Note that $\hat{i} \cdot \hat{i} = \hat{j} \cdot \hat{j} = 1$ (1.7)

$\qquad \hat{i} \cdot \hat{j} = \hat{j} \cdot \hat{i} = 0$ (1.8)

A vector product is a vector together with size and direction. Size is the area of a rectangle made by the vectors $\vec{A}$ and $\vec{B}$. Direction $\vec{R}$ is a thumb point when one turns finger from plane $\vec{A}$ to $\vec{B}$ according to the right-hand rule, as shown in Fig. 1.3.

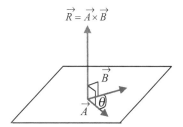

Figure 1.3   Vector products.

From Fig. 1.3, we get

$$\left| \vec{A} \times \vec{B} \right| = \left| \vec{A} \right| \left| \vec{B} \right| \sin\theta$$ (1.9)

where θ is the angle between the vectors $\vec{A}$ and $\vec{B}$.

Note that $\hat{i} \times \hat{i} = \hat{j} \times \hat{j} = 0$ (1.10)

$\qquad \hat{i} \times \hat{j} = \hat{k} \ \ (\text{or } \hat{j} \times \hat{i} = -\hat{k})$ (1.11)

$\qquad \hat{j} \times \hat{k} = \hat{i}$ (1.12)

$\qquad \hat{k} \times \hat{i} = \hat{j}$ (1.13)

In addition to the doublet vectors ($\vec{A}$ and $\vec{B}$), there is also a product of three vectors $\vec{A}, \vec{B},$ and $\vec{C}$, which needs to pay attention to the ordering of vectors $\vec{A}, \vec{B},$ and $\vec{C}$ as follows:

$$\vec{A} \cdot (\vec{B} \times \vec{C}) = (\vec{A} \times \vec{B}) \cdot \vec{C} = (\vec{C} \times \vec{A}) \cdot \vec{B} \tag{1.14}$$

$$\vec{A} \times (\vec{B} \times \vec{C}) = \vec{B}(\vec{A} \cdot \vec{C}) = \vec{C}(\vec{A} \cdot \vec{B}) \tag{1.15}$$

$$\vec{A} \times (\vec{B} \times \vec{C}) \neq (\vec{A} \times \vec{B}) \times \vec{C} \tag{1.16}$$

1.4 Gradient, Divergence, and Rotation of Vectors

First we define the length of an area vector as the area of surface S surrounded by a closed loop. Its direction is defined as the direction normal to the plane S, as shown in Fig. 1.4.

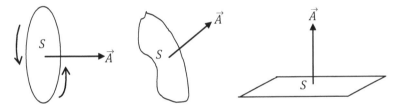

Figure 1.4 Area vector for a disc (left), an arbitrary two-dimensional shape (middle), and a rectangle (right).

The definition of direction of these vectors is dependent on cases given by

Nabla or Del $\quad \vec{\nabla} \equiv \left(\dfrac{\partial}{\partial x}, \dfrac{\partial}{\partial y}, \dfrac{\partial}{\partial z} \right) \tag{1.17}$

The gradient of a scalar is the increasing rate of the scalar quantity given by

$$\vec{\nabla}\varphi = \text{grad } \varphi = \left(\dfrac{\partial \varphi}{\partial x}, \dfrac{\partial \varphi}{\partial y}, \dfrac{\partial \varphi}{\partial z} \right) \tag{1.18}$$

Where $\varphi(x, y, z)$ is a scalar field.

The divergence of a vector is the net flow of vector that springs out from or absorbed into a certain point and is given by

$$\vec{\nabla} \cdot \vec{A} = div\ \vec{A} = \frac{\partial A_x}{\partial x} + \frac{\partial A_y}{\partial y} + \frac{\partial A_z}{\partial z} \tag{1.19}$$

where $\vec{A}(x,y,z)$ is a vector field.

The rotation of a vector is the quantity of a vector that rotates and is given by

$$\vec{\nabla} \times \vec{A} = rot\ \vec{A} = \begin{pmatrix} \dfrac{\partial A_z}{\partial y} - \dfrac{\partial A_y}{\partial z} \\[2mm] \dfrac{\partial A_x}{\partial z} - \dfrac{\partial A_z}{\partial x} \\[2mm] \dfrac{\partial A_y}{\partial x} - \dfrac{\partial A_x}{\partial y} \end{pmatrix} \tag{1.20}$$

In addition, there is an important notation given by

$$\text{Laplacian}\quad \Delta = \vec{\nabla} \cdot \vec{\nabla} = \vec{\nabla}^2 \equiv \frac{\partial^2}{\partial x^2} + \frac{\partial^2}{\partial y^2} + \frac{\partial^2}{\partial z^2} \tag{1.21}$$

1.5 Gauss's and Stokes' Theorem

According to Gauss's theorem, the total quantity of divergence of vectors in a closed volume V (Fig. 1.5) is the sum of the vectors going out across a boundary of the volume given by

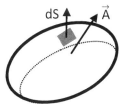

Figure 1.5 Geometry for Gauss's theorem.

$$\int_V \vec{\nabla} \cdot \vec{A}\, dV = \int_s \vec{A} \cdot \vec{n}\, dS \tag{1.22}$$

According to Stokes' theorem, the total quantity of rotation of vortices of vectors in a finite plane is equivalent to the rotation of the vector on the boundary of the plane given by (Fig. 1.6)

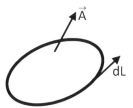

Figure 1.6 Geometry of Stokes' theorem.

$$\int_S (\vec{\nabla} \times \vec{A}) \cdot \vec{n} \, dS = \int_C \vec{A} \cdot \vec{t} \, dL \qquad (1.23)$$

1.6 Maxwell's Equations

All electromagnetic phenomena, including relativistic effects, can be described by Maxwell's equations, when the interactions between the charged particles and the electromagnetic fields, i.e., Lorentz force, are not involved. Maxwell's equations are displayed by the combination of the following four equations:

Faraday's law: $\vec{\nabla} \times \vec{E} = -\dot{\vec{B}}$ (1.24)

where $\dot{\vec{B}} = \dfrac{\partial \vec{B}}{\partial t}$

Ampere's law: $\vec{\nabla} \times \vec{H} = \vec{J} + \dot{\vec{D}}$ (1.25)

where $\dot{\vec{D}} = \dfrac{\partial \vec{D}}{\partial t}$

Gauss's law for electrostatics: $\vec{\nabla} \cdot \vec{D} = \rho_{\text{TRUE}}$ (1.26)

Gauss's law for magnetostatics: $\vec{\nabla} \cdot \vec{B} = 0$ (1.27)

Here $\vec{E}, \vec{D}, \vec{H}$, and $\vec{B}$ are the electric field strength, the electric displacement (the flux density), the magnetic field strength, and the magnetic flux density, respectively.

The physical interpretation of Eq. 1.24, called Faraday's law of electromagnetic induction, is that a temporal variation in magnetic flux density induces a spiral electric field, as shown in Fig. 1.7.

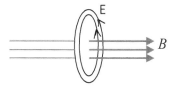

Figure 1.7   Physical interpretation of Faraday's law.

The electric field is induced to cancel the effect of the variation in the magnetic field by electric current.

The physical interpretation of Eq. 1.25, called Ampere's law of electromagnetic induction, is that electric current is the displacement current that induces a magnetic field loop, as shown in Fig. 1.8.

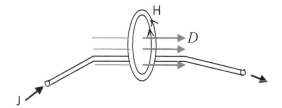

Figure 1.8   Physical interpretation of Ampere's law.

The physical interpretation of Eq. 1.26, called Gauss's law of electric field, is shown in Fig. 1.9.

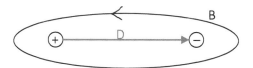

Figure 1.9   Physical interpretation of Gauss's law.

Electric displacement gushes out from electric charges. The physical interpretation of Eq. 1.27, called Gauss's law of magnetic field, is also shown in Fig. 1.9. There is no gushing of magnetic field from anywhere.

1.7   Problems

1. Show some examples of scalar and vector quantities.

2. Prove that $\vec{A} \cdot \vec{B} = \vec{B} \cdot \vec{A}$

3. Prove that $\vec{A} \times \vec{A} = 0$.

4. Prove that $\vec{A} \times \vec{B} = -\vec{B} \times \vec{A}$.

5. Prove that $\vec{A} \times (\vec{B} \times \vec{C}) \neq (\vec{A} \times \vec{B}) \times \vec{C}$.

6. Explain the concepts of Laplacian and Nabla.

7. Discuss the difference between divergence and rotation of vectors.

8. Write Gauss's theorem and Stokes' theorem.

9. Write Maxwell's equations and discuss the physical meanings of Maxwell's equations.

10. Discuss the Lorentz force describing the force between electromagnetic field and a charged particle.

Chapter 2

Electromagnetic Units and Electric Charges

2.1 Introduction

In this chapter, we will study the units used in electromagnetic theory. Learning all electromagnetism units is considered difficult because there are several systems of units. So here we will learn only the MKS system. In this chapter, we will also learn about the forces acting between electric charges.

2.2 MKS Units

A concept that electromagnetic theory has and classical dynamics does not is charge. In MKS units, when there is a Coulomb force of 9×10^9 N between two equal charges 1 m apart from each other, we can define a charge of 1 coulomb (1 C). Moreover, when a charge of 1 C passes through at a given point of interest in 1 s, we can define that current as 1 ampere (1 A). All physical quantities can be written by a combination of the following four units.

For calculating a physical quantity in the electromagnetic theory, we always need to attach units to the result. If we do not know the unit of a physical quantity, we can find the unit in the following way. First, write an equation consisting of the physical quantity in

Optical Properties of Solids: An Introductory Textbook
Kitsakorn Locharoenrat
Copyright © 2016 Pan Stanford Publishing Pte. Ltd.
ISBN 978-981-4669-06-1 (Hardcover), 978-981-4669-07-8 (eBook)
www.panstanford.com

question. Second, attach units to every physical quantity appearing in the question. Third, from the equality of the product of units on both sides of the equation, we can find the unit that should be attached to the physical quantity in question.

Table 2.1 Examples of basic MKS units

Quantity	Unit
Length	1 meter (1 m)
Weight	1 kilogram (1 kg)
Time	1 second (1 s)
Electric current	1 ampere (1 A)

Maxwell's equations presented in Chapter 1 are written in MKS units. The units attached to the electromagnetic quantities appearing in Maxwell's equations can be written as in Table 2.2. Be sure to attach units to every numerical value, not only at the end but also during calculations.

Table 2.2 Electromagnetic quantities

Quantity	Symbol	Unit
Electric field strength	$\vec{E}$	V/m
Electric displacement	$\vec{D}$	C/m^2
Magnetic field strength	$\vec{H}$	A/m
Magnetic flux density	$\vec{B}$	$T = V \times s \times m^{-2}$
Current density	$\vec{J}$	A/m^2
Charge density	ρ	C/m^3
div, rot	$\vec{\nabla}\cdot, \vec{\nabla}\times$	m^{-1}
Differential with respect to time	d/dt	s^{-1}
Electrical resistivity	R	Ω
Electrical capacity	C	F
Inductance	L	H
Dielectric constant	ε_0	8.854×10^{-12} F/m
Magnetic permeability	μ_0	$4\pi \times 10^{-7}$ H/m

Quantity	Symbol	Unit
Magnetization (E–B correspondent)	$\vec{M}$	A/m
Magnetization (E–H correspondent)	$\vec{P}_M$	Wb/m^2

Nowadays scientific papers are not accepted unless equations are written in MKS units. However, past papers were written in emu, esu, or gauss units. Hence we must also learn how to convert these units. Particularly those who study magnetism cannot understand these papers unless they understand emu units.

The relationship between electric displacement $\vec{D}$ and electric field strength $\vec{E}$ can be written as

$$\vec{D} = \varepsilon\,\vec{E} \tag{2.1}$$

One the other hand, the relationship between magnetic flux density $\vec{B}$ and magnetic field strength $\vec{H}$ can be expressed as

$$\vec{B} = \mu\,\vec{H} \tag{2.2}$$

Here ε and μ are tensors that present the electric and magnetic responses of materials (e.g., dielectrics and magnetic materials), respectively.

2.3 Coulomb's Law

Electric charges are discovered in triboelectricity when two things are rubbed against each other. Table 2.3 represents triboelectric series. Since atoms consist of nucleus and electrons, they are charged with integral multiples of the elementary electric charge of 1.602×10^{-19} C. The total amount of the charges is conserved.

Table 2.3   Triboelectric series

Asbestos	Increasingly positive (+)
Glass	
Mica	
Wool	
Aluminum	
Nickel, copper	
Brass, silver	
Celluloid	Increasingly negative (−)

French physicist Coulomb experimentally measured the electrostatic force between two electric charges. He founded that the electric force is inversely proportional to the square of the gap of the charges, however, it is directly proportional to the multiplication of the charges.

From Fig 2.1, Coulomb introduced Coulomb's law stating that the force between two electric charges q_1 and q_2, with r being the space between the two charges is

$$\vec{F} = \frac{1}{4\pi\varepsilon_0} \cdot \frac{q_1 q_2}{r^2} \cdot \frac{\vec{r}}{r} \quad \text{in vacuum} \tag{2.3}$$

$$\vec{F} = \frac{1}{4\pi\varepsilon} \cdot \frac{q_1 q_2}{r^2} \cdot \frac{\vec{r}}{r} \quad \text{in dielectric} \tag{2.4}$$

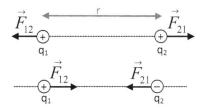

Figure 2.1   Forces between electric charges.

When q_1 and q_2 are of the same sign, the force $\vec{F}$ is a repulsive force. When q_1 and q_2 are of opposite signs, $\vec{F}$ is an attractive force. Here ε_0 is known as the permittivity of the vacuum, and ε is the permittivity of the dielectric. The following equality is convenient to remember:

$$\frac{1}{4\pi\varepsilon_0} = 9 \times 10^9 \text{ m} \cdot \text{F}^{-1} \tag{2.5}$$

When many charges distribute, the Coulomb force applied to one charge is the vector sum of the Coulomb forces between this charge and the other charges.

2.4   Conductor and Insulator

When a material is charged and the charge can move within the material, that material is known as a conductor (Fig. 2.2a). A charged conductor completely discharges when grounded.

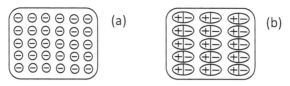

Figure 2.2   Type of materials: (a) conductor and (b) insulator.

On the other hand, when the charge cannot move within the material, the material is known as an insulator (Fig. 2.2b). Grounding insulator neither removes nor prevents surface charge.

When charge distribution varies in a conductor partly because of the external electric field, the phenomenon is called *electrostatic induction*.

2.5   Problems

1. Write the units of the following electromagnetic quantities in MKS units: (a) electric resistivity, (b) inductance, and (c) capacitance.
2. Discuss the differences between CGS and MKS units.
3. Discuss the relation of electric charges between fur and amber.
4. Discuss the Coulomb force between two charges.
5. Explain why the total amount of charges is conserved.
6. Explain the relation between dielectric constant and magnetic permeability.
7. Write the Coulomb's law for three charges in vacuum.
8. Write the relation between Coulomb constant and dielectric constant.
9. Discuss the differences between induction and conduction.
10. Discuss the differences between a conductor, insulator, and semiconductor.

Chapter 3

Electric Field and Electric Potential

3.1 Introduction

Instruments such as electron microscope, electron monochromator, and ion monochromator operate under the principle of electric fields. Hence, as the first step toward understanding these instruments, it is important to study electric fields and electric potential partly due to electric charges. This chapter deals with electric fields and electric potential due to electric charges, Gauss's law, and electric dipoles. It also presents calculation methods for several electric field distributions.

3.2 Electric Field

The electric field is the field of the electric force that a charge receives. The force $\vec{F}$ acting on a charge q_1 under an electric field $\vec{E}$ (Fig. 3.1) is given by

$$\vec{F} = q_1 \vec{E} \qquad (3.1)$$

The electric field $\vec{E}$ emerging from another charge q_2 at a distance r from the charge q_1 is given by

$$\vec{E} = \frac{1}{4\pi\varepsilon} \frac{q_2}{r^2} \cdot \frac{\vec{r}}{r} \qquad (3.2)$$

Optical Properties of Solids: An Introductory Textbook
Kitsakorn Locharoenrat
Copyright © 2016 Pan Stanford Publishing Pte. Ltd.
ISBN 978-981-4669-06-1 (Hardcover), 978-981-4669-07-8 (eBook)
www.panstanford.com

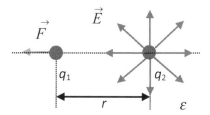

Figure 3.1   Forces between charges q_1 and q_2.

Combining Eqs. 3.1 and 3.2, we get

$$\vec{F} = \frac{1}{4\pi\varepsilon}\frac{q_1 q_2}{r^2} \cdot \frac{\vec{r}}{r} \tag{3.3}$$

This is Coulomb's law we learned in Chapter 2. The Coulomb force in Eq. 3.3 can be understood as the force acting on the charge q_1 under the electric field due to the other charge q_2. We normally understand all electromagnetic effects by using the concept of fields. From Eqs. 3.1 and 3.2, we assume that the space is filled with *a uniform dielectric*. If we apply these equations to vacuum, we can put ε in place of ε_0.

3.3   Lines of Electric Force

Lines of electric force are virtual lines with directional arrows drawn in an electric field, as shown in Fig. 3.2.

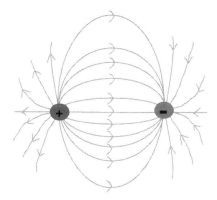

Figure 3.2   Field lines (lines of force) at dipole charges.

A tangent to a line of electric force corresponds to the direction of the electric field at that point. Lines of electric force of high density indicate a strong electric field. As we limit ourselves to a static case, we see that the line of electric force begins at the positive charge and finishes at the negative charge. It never disappears where there is no charge. The number of lines coming out of a charge depends on the quantity of that charge according to Gauss's law. The lines of electric force are always perpendicular to the conductor surface.

3.4 Electric Potential

The work done on a charge q_1, when the charge is moved from location A to location B receiving a force $\vec{F}$ from an electric field $\vec{E}$, becomes the kinetic energy W of this charge and is given by

$$W = \int_A^B -\vec{F} \cdot d\vec{s} \tag{3.4}$$

$$W = -\int_A^B q_1 \vec{E} \cdot d\vec{s} \tag{3.5}$$

$$W = q_1 V_{BA} \tag{3.6}$$

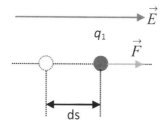

Figure 3.3   Electric potential due to a point charge q_1.

Here, V_{BA} is known as the electric potential (difference). We define V_{BA} as

$$V_{BA} = -\int_A^B \vec{E} \cdot d\vec{s} \tag{3.7}$$

A surface with a constant electric potential is known as an equipotential surface. If the electric potential is compared to a geometrical height, the electric field can be compared to the gradient of a geometrical shape.

We use the symbol V to indicate the electric potential when we consider the potential difference. On the other hand, we use the symbol ϕ when we consider the absolute potential with $\phi = 0$ at infinity, as shown in Fig. 3.4. The equipotential surface and the lines of electric force are always perpendicular to each other. Starting from Eq. 3.4, we have

$$W = \int_A^B -\vec{F} \cdot d\vec{s}$$

$$\text{or} \quad W = -\int_A^B \frac{1}{4\pi\varepsilon} \frac{q_1 q_2}{r^2} \frac{\vec{r}}{r} \cdot d\vec{s} \tag{3.8}$$

$$W = -\int_\infty^r \frac{1}{4\pi\varepsilon} \frac{q_1 q_2}{r^2} \cdot dr \tag{3.9}$$

$$W = \frac{q_1 q_2}{4\pi\varepsilon} \left[\frac{1}{r} \right]_\infty^r = \frac{q_1 q_2}{4\pi\varepsilon r} = q_2 \phi \tag{3.10}$$

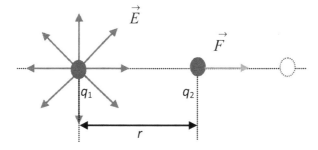

Figure 3.4   Electric potential due to a point charge q_1.

With $\phi = 0$ at infinity, the electric potential due to a point charge q_1 is

$$\phi = \frac{q_1}{4\pi\varepsilon r} \tag{3.11}$$

In the case of several charges, it is convenient to use the principle of superposition when we calculate the electric potential.

By differentiating Eq. 3.7, we find that the electric field strength is negative of the gradient of the electric potential:

$$\vec{E} = -\vec{\nabla} V_{BA} \qquad (3.12)$$

3.5 Gauss's Law

Suppose a closed surface S (area = $4\pi r^2$) is filled with a uniform dielectric in an electric field. Gauss's law has a benefit in calculating the electric field because of the charges. It can be proved by using the third Maxwell's equation (Eq. 1.26):

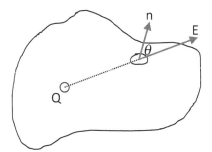

Figure 3.5 Gaussian surface S with a point charge within it.

$$\vec{\nabla} \cdot \vec{D} = \rho_{Q_{TRUE}}$$

$$\int_V \vec{\nabla} \cdot \vec{D}\, dV = \int_V \rho_{Q_{TRUE}}\, dV \qquad (3.13)$$

and Gauss's theorem (Eq. 1.22):

$$\int_V \vec{\nabla} \cdot \vec{D}\, dV = \int_S \varepsilon \vec{E} \cdot \vec{n}\, dS$$

where ε is the dielectric constant of the medium.

Substituting Eq. 1.22 in Eq. 3.13, we get

$$\int_S \vec{E} \cdot \vec{n}\, dS = \frac{1}{\varepsilon} Q_{TRUE} \qquad (3.14)$$

The electric displacement integrated over this surface S is equal to the true charge inside the surface as

$$\int_S \vec{D} \cdot \vec{n} \, dS = Q_{\text{TRUE}} \qquad (3.15)$$

We find that the number of electric force lines running through the surface S is $1/\varepsilon$ times the total true charge inside the surface. Suppose a certain amount of hot water comes out of a hot spring. The total amount of hot water flowing out of a district with hot springs is dependent on the total number of hot springs inside the district, unless the water is absorbed by the ground. This story is similar to the phenomenon of Gauss's law discussed above. Here, a charge has been compared to a hot spring, and the number of lines of electric force has been compared to the amount of hot water.

Finally, by using Eq. 3.14, we can prove Eq. 3.2:

$$\vec{E} = \frac{1}{4\pi\varepsilon} \frac{q_2}{r^2} \cdot \frac{\vec{r}}{r}$$

3.6   Ohm's Law

Ohm's law represents the relation between current I and potential difference V_{BA} measured across a conductor of length l with cross-sectional area S by starting from Eq. 3.7 (Fig. 3.6):

$$V_{\text{BA}} = -\int_A^B \vec{E} \cdot d\vec{s} = -E\,l$$

$$V_{\text{BA}} = -\frac{J}{\sigma} l = -\frac{(I/S)}{\sigma} l \qquad (3.16)$$

where σ is conductivity.

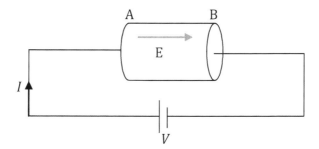

Figure 3.6   Principle of Ohm's law.

If we define resistance as

$$R = \frac{l}{\sigma S}$$
(3.17)

Eq. 3.16 becomes

$$V_{BA} = -I R$$
(3.18)

In Ohm's law, there can be no electric field in a conductor. In addition, there can be no charge within a conductor; however, charges can exist only on the surface of the conductor.

3.7 Electric Dipole

In addition to a point charge, when two charges $+q$ and $-q$ are located very close to each other, they form an electric dipole (Fig. 3.7).

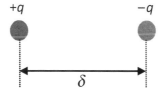

Figure 3.7   Electric dipole.

If the distance between these charges is δ, $q\delta$ is the moment of that electric dipole. Polarization in a dielectric (Chapter 5) consists of electric dipoles. Radiation of absorption of electromagnetic waves by atoms or molecules occurs as a result of interaction between electric dipole and radiation field. Therefore, the concept of electric dipole is very important. The electric potential ϕ, in polar coordinates as in Eq. 3.11, due to an electric dipole with the electric dipole moment $\vec{p} = q\delta$ is modified as

$$\phi = \frac{1}{4\pi\varepsilon}\frac{q\delta\cos\theta}{r^2}$$
(3.19)

Differentiating Eq. 3.19 with respect to the polar coordinates (Fig. 3.8), we can calculate the electric field strength due to the electric dipole as follows.

From Eq. 3.12, we have $\vec{E} = -\vec{\nabla}\phi$

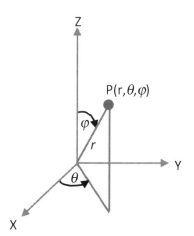

Figure 3.8   Electric field strength due to the electric dipole with polar coordinates (r, θ, φ).

$$\vec{E} = E_r + E_\theta + E_\varphi \tag{3.20}$$

$$\vec{E} = \left[-\vec{\nabla}\phi \right]_r + \left[-\vec{\nabla}\phi \right]_\theta + \left[-\vec{\nabla}\phi \right]_\varphi \tag{3.21}$$

$$\vec{E} = \frac{-\partial\phi}{\partial r} - \frac{1}{r}\frac{\partial\phi}{\partial\theta} - \frac{1}{r\sin\theta}\frac{\partial\phi}{\partial\varphi} + 0 \tag{3.22}$$

$$\vec{E} = \frac{-\partial}{\partial r}\left(\frac{1}{4\pi\varepsilon} \cdot \frac{q\delta\cos\theta}{r^2} \right) \frac{-1}{r}\frac{\partial}{\partial\theta}\left(\frac{1}{4\pi\varepsilon}\frac{q\delta\cos\theta}{r^2} \right) + 0 \tag{3.23}$$

$$\vec{E} = \frac{2}{4\pi\varepsilon} \cdot \frac{q\delta\cos\theta}{r^3} + \frac{1}{4\pi\varepsilon}\frac{q\delta\sin\theta}{r^3} + 0 \tag{3.24}$$

On the other hand, for an electric dipole layer as displayed in Fig. 3.9, electric potential in Eq. 3.11 due to an electric dipole layer is modified as

$$\phi = \frac{M}{4\pi\varepsilon}\Omega \tag{3.25}$$

Here, M is the surface–dipole layer density equal to $\sigma\delta$ (C/m). Ω is the solid angle subtended at the observation point by the dipole layer in question. When the observer goes far away from the dipole layer, Ω becomes small and ϕ approaches zero.

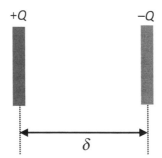

+Q   −Q

δ

Figure 3.9   Electric dipole layer.

3.8   Problems

1. Prove Gauss's law.
2. Derive the electric field because of a point charge by using Gauss's law.
3. Discuss Ohm's law.
4. Prove that the electric field strength is negative of the gradient of the electric potential.
5. Derive the electric potential ϕ due to a point charge:

$$\phi = \frac{q_1}{4\pi\varepsilon r}$$

6. Derive the electric potential ϕ due to an electric dipole:

$$\phi = \frac{1}{4\pi\varepsilon}\frac{q\delta\cos\theta}{r^2}$$

7. Derive the electric potential ϕ due to an electric dipole layer:

$$\phi = \frac{M}{4\pi\varepsilon}\Omega$$

8. Derive the electric field at the space r from the infinite line charge with the uniform charge density per length $\tau\,(\text{C/m})$.
9. Derive the electric field at the space r from an infinite surface charge together with the uniform charge density per area σ (C/m^2).
10. Derive the electric field at the space r from an infinite volume charge together with the uniform charge density per volume $\rho\,(\text{C/m}^3)$.

Chapter 4

Capacitance and Electromagnetic Energy

4.1 Introduction

Impedance, capacitance, and inductance are the most frequent concepts studied in electronics. In fact we sometimes experience being a capacitor in our daily life. In a dry, cold day, when we suddenly get an electric shock on our fingers when we touch a metal door handle, our body actually functions as a capacitor and releases electricity to the door handle. This chapter deals with capacitance and capacitor, which store electricity. It also presents the concept of electromagnetic theory.

4.2 Capacitance

A system in which two conductors are spaced out by an insulator is called a *capacitor*. When we put charges of opposite signs on these two conductors, the charges attract each other and get stabilized. However, when we put charges on an isolated conductor, the charges do not know where to go and stay on the conductor. These conductor systems are capable of storing electricity and are said to have capacitances.

When charges Q are accumulated on conductors, electric fields emerge because of these charges. Thus, the conductors have electric

Optical Properties of Solids: An Introductory Textbook
Kitsakorn Locharoenrat
Copyright © 2016 Pan Stanford Publishing Pte. Ltd.
ISBN 978-981-4669-06-1 (Hardcover), 978-981-4669-07-8 (eBook)
www.panstanford.com

potential V. The potential V, capacitance C, and charges Q have the following relation:

$$Q = CV \qquad (4.1)$$

Capacitance is just like the area of the bottom of a cup, as shown in Fig. 4.1. The volume of water (or Q) is this area (or C) multiplied by the height of water (V). The unit of capacitance is farad (F).

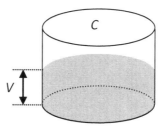

Figure 4.1   Model of concept of capacitance.

4.3   Electromagnetic Energy

A charged capacitor naturally has electromagnetic energy as

$$W = \frac{1}{2}QV \qquad (4.2)$$

or

$$W = \frac{1}{2}CV^2 = \frac{1}{2}\frac{Q^2}{C} \qquad (4.3)$$

This relation can be proved in several ways—by calculating (i) Joule's heat generated by this charge, (ii) the energy of a charged capacitor by gradually increasing the distance between the two electrodes, and (iii) the electromagnetic field energy stored under the space between the two electrodes.

4.3.1   Calculating Joule's Heat Generated by Charges

Joule's heat generated by the charges can be calculated with the help of Fig. 4.2.

Starting from $I = \frac{1}{2}\frac{V}{R}$ $\qquad (4.4)$

$$t = \frac{Q}{I} = \frac{Q}{V/2R} \tag{4.5}$$

$$W = Pt = I^2\,Rt \tag{4.6}$$

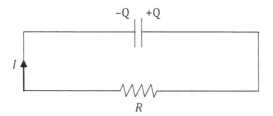

Figure 4.2   RC circuit.

Substituting Eqs. 4.4 and 4.5 in Eq. 4.6, we get

$$W = R \cdot \left(\frac{V}{2R}\right)^2 \left(\frac{Q}{V/2R}\right) \tag{4.7}$$

$$W = \frac{1}{2}QV \qquad\qquad \text{(which is Eq. 4.2)}$$

4.3.2   Calculating Energy of Charged Capacitor

The energy of a charged capacitor can be calculated by gradually increasing the distance between the two electrodes (if $W \neq 0$), as exhibited in Fig. 4.3.

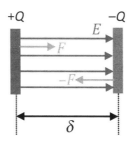

Figure 4.3   A conductor system.

We know that $F = \dfrac{1}{2}QE$ \hfill (4.8)

$$W = F\delta = \frac{1}{2}Q\,(E\delta) \tag{4.9}$$

Thus, a charged capacitor naturally has electromagnetic energy as

$$W = \frac{1}{2}QV \qquad \text{(which is Eq. 4.2)}$$

4.3.3 Calculating Electromagnetic Field Energy Stored under the Space between Two Electrodes

We calculate the electromagnetic field energy density W stored under the space between two electrodes as

$$W = \int_V \left(\frac{1}{2}\varepsilon E^2 + \frac{1}{2}\mu H^2 \right) dV \cong \frac{1}{2}\varepsilon E^2 V \qquad (4.10)$$

$$W = \frac{1}{2}\varepsilon E^2 (S\delta) = \frac{1}{2}\varepsilon ES\, E\delta \qquad (4.11)$$

From Eq. 3.14, we have

$$\int_S \vec{E} \cdot \vec{n}\, dS = \frac{1}{\varepsilon}Q_{\text{total}}$$

or $\quad ES = \dfrac{Q}{\varepsilon} \quad \Rightarrow \quad Q = \varepsilon E\, S \qquad (4.12)$

Substituting Eq. 4.12 in Eq. 4.11, we finally obtain

$$W = \frac{1}{2}QV \qquad \text{(which is Eq. 4.2)}$$

We may be annoyed by the factor 0.5 in Eq. 4.2 when we learn the potential energy of a point charge in an electric field:

$$W = qV \qquad \text{(which is Eq. 3.6)}$$

This is neither a mystery nor a contradiction. The electric field and the electric potential in a capacitor are produced by the charge itself and are zero when there is no stored charge. In the case of a point charge under an electric field, the electric field exists no matter whether the charge in question exists or not. This difference of the energy between a charge in free space and this charge under two electrodes is reflected as the difference of factor 0.5.

4.4 Coefficient of Electrostatic Potential

Suppose that there exists a number of conductors. A relation holds between the charge Q_i on the ith conductor and the electric potential V_j of the jth conductor:

$$Q_i = \sum_j C_{ij} V_j \tag{4.13}$$

$$V_j = \sum_i P_{ji} Q_i \tag{4.14}$$

Here P_{ji} is known as the coefficient of electrostatic potential and C_{ij} is the capacitance. If i is not equivalent to j, C_{ij} is sometimes called the coefficient of induction. From Eqs. 4.13 and 4.14, we find the following:

1. C_{ij} and P_{ji} are inverse matrices of each other.
2. A reciprocity $C_{ij} = C_{ji}$ and $P_{ij} = P_{ji}$ holds.
3. The electromagnetic energy of the conductor system is

$$W = \frac{1}{2}\sum_i Q_i V_i = \frac{1}{2}\sum_{ij} P_{ij} Q_i Q_j = \frac{1}{2}\sum_{ij} C_{ij} V_i V_j \tag{4.15}$$

For calculating the capacitance of a conductor system, we calculate the electric field and the electric potential by using Gauss's law (Eq. 3.14) after putting some charge on the conductors and then using the relation $C = Q/V$. Furthermore, it is convenient to assume that the electric potential becomes zero at infinity.

4.5 Problems

1. Prove that the potential energy of a point charge in an electric field is

   $W = qV$

2. Prove that a charged capacitor naturally has electromagnetic energy as

   $$W = \frac{1}{2}QV$$

3. For an RC circuit, prove that

   $$I = \frac{1}{2}\frac{V}{R}$$

4. For a conductor system, prove that

   $$F = \frac{1}{2}QE$$

5. Prove that

$$W \cong \frac{1}{2}\varepsilon E^2 V$$

6. Discuss the difference between the electromagnetic energies of a point charge and a capacitor.
7. Calculate the capacitance of a metal sphere of radius a (m) in vacuum.
8. Calculate the capacitance of concentric double metal spheres of radii a (m) and b (m) in vacuum in which a is less than b. Assume that one of the spheres is grounded.
9. Calculate the capacitance per unit length of coaxial double metal cylinders of radii a (m) and b (m) in vacuum in which a is less than b. Assume that one of the spheres is grounded.
10. Calculate the coefficient of electrostatic potential of three concentric metal spheres of radii a_1, a_2, and a_3 ($a_1 < a_2 < a_3$). Assume that none of the spheres is grounded.

Chapter 5

Dielectric Materials

5.1   Introduction

Materials except magnetic materials can be regarded as dielectrics in electromagnetic theory and optics. Dielectric (or dielectric material) is a nonconducting material, but it is an efficient supporter of the electric field. We can say without exaggerating that their dielectric constant describes all their macroscopic properties. Thus, we should thoroughly study dielectric constants before we learn solid-state physics. In this chapter, we will learn how to calculate the electric field in charge distributions. If dielectrics are involved in this calculation, it may be difficult to solve. However, it would be easy if we properly understand the concept of dielectric displacement and Maxwell's boundary conditions.

5.2   Polarization

Polarization is the displacement of charges when we put a dielectric in an electric field. Under the electric field, $\vec{E}$ the molecules or units of cells in the dielectric are electrically polarized, and a macroscopic polarization $\vec{P}$ emerges, as shown in Fig. 5.1. Quite often polarization is parallel to the applied electric field and can be pointed in the same direction. A polarized dielectric has polarized charges on its

Optical Properties of Solids: An Introductory Textbook
Kitsakorn Locharoenrat
Copyright © 2016 Pan Stanford Publishing Pte. Ltd.
ISBN 978-981-4669-06-1 (Hardcover), 978-981-4669-07-8 (eBook)
www.panstanford.com

both ends. Between polarization $\vec{P}$ and electric field $\vec{E}$, we get the following relation:

$$\vec{P} = \chi_e \vec{E} \tag{5.1}$$

where χ_e represents the electric susceptibility of the medium and has the same dimension as the dielectric constant. The macroscopic polarization $\vec{P}$ is related to the microscopic electric dipole moment $\vec{p}$ through the following relation:

$$\vec{P} = N\vec{p} \tag{5.2}$$

where N is the number of atoms within a unit volume.

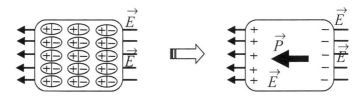

Figure 5.1 Interaction of electromagnetic field with a dielectric.

5.3 Dielectric Displacement and Dielectric Constant

Let us define the dielectric displacement $\vec{D}$ as

$$\vec{D} = \varepsilon_0 \vec{E} + \vec{P} \tag{5.3}$$

Substituting Eq. 5.1 in Eq. 5.3, we get

$$\vec{D} = (\varepsilon_0 + \chi_e)\vec{E} \tag{5.4}$$

If ε is the *dielectric constant*, it is written as

$$\varepsilon = \varepsilon_0 + \chi_e \tag{5.5}$$

Substituting Eq. 5.5 in Eq. 5.4, we get

$$\vec{D} = \varepsilon \vec{E} \tag{5.6}$$

$$\varepsilon = \varepsilon_0(1 + \overline{\chi_e}) \tag{5.7}$$

$$\varepsilon = \varepsilon_0 \varepsilon_s \tag{5.8}$$

where ε_s and $\overline{\chi_e}$ are relative dielectric constant and relative electric susceptibility, respectively. They are obtained by dividing ε and χ_e by ε_0, respectively.

Between a relative dielectric constant $\varepsilon_s(\omega)$ and a refractive index $n(\omega)$ holds a relation

$$\varepsilon_s(\omega) = n^2(\omega) \tag{5.9}$$

The capacitance of a capacitor with a dielectric inserted is

$$C_i = \varepsilon_s C_0 \tag{5.10}$$

By taking divergence of Eq. 5.3, we obtain the relation

$$\vec{\nabla} \cdot \vec{D} = \varepsilon_0 \vec{\nabla} \cdot \vec{E} + \vec{\nabla} \cdot \vec{P} \tag{5.11}$$

We rewrite Eq. 5.11 as follows:

$$\rho_{\text{TRUE}} = \rho_{\text{TOTAL}} - \rho_p \tag{5.12}$$

Here, we define the following relations:

True charge $\quad \rho_{\text{TRUE}} = \vec{\nabla} \cdot \vec{D}$ \hfill (5.13)

Total charge $\quad \rho_{\text{TOTAL}} = \varepsilon_0 \vec{\nabla} \cdot \vec{E}$ \hfill (5.14)

Polarization charge $\quad \rho_p = -\vec{\nabla} \cdot \vec{P}$ \hfill (5.15)

Namely, $\vec{D}$, $\vec{E}$, and $\vec{P}$ are the fields coming out from the true, total, and polarization charges, respectively. For each field and charge, Gauss's law holds. The term "TRUE" in the concept of "true charge" means that it does not include polarization charge. So if we add the polarization charge to the true charge, we call it "total charge." As seen in Fig. 5.1, the polarization vector gathers at the positive polarization charge. Then the polarization charge has a sign opposite to that of the divergence of the polarization vector.

5.4 Maxwell's Boundary Conditions

A boundary condition is useful when we know a field in one surface and want to find it in another location. Here we present four Maxwell's boundary conditions. When an electric field crosses a surface, as seen in Fig. 5.2, the first boundary condition is

$$\vec{E}_{1//} = \vec{E}_{2//} \tag{5.16}$$

From the relation $\vec{D} = \varepsilon \vec{E}$, the second boundary condition is

$$\vec{D}_{1\perp} - \vec{D}_{2\perp} = \rho_s \tag{5.17}$$

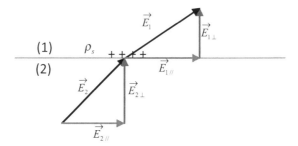

Figure 5.2   Electric field crosses from one region to another.

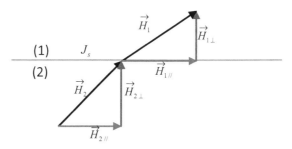

Figure 5.3   Magnetic field crosses from one location to another.

When a magnetic field crosses a surface, as seen in Fig. 5.3, the third boundary condition is

$$\vec{H}_{1//} - \vec{H}_{2//} = \vec{J}_s \tag{5.18}$$

From the relation $\vec{B} = \mu\,\vec{H}$, the fourth boundary condition is

$$\vec{B}_{1\perp} = \vec{B}_{2\perp} \tag{5.19}$$

5.5   Problems

1. Discuss the electric susceptibility of a medium related to the polarization field.
2. Discuss the relation between electric dipole and polarization.
3. Discuss the difference between true charge and total charge.
4. Discuss the physical meaning of electric displacement.
5. Discuss the relation between dielectric constant and refractive index.

6. Explain the relation between the capacitance of a capacitor and dielectric constant.

7. Suppose that a dielectric sphere of dielectric constant ε and radius a is given a uniform charge of charge density ρ. Calculate the electric potential everywhere.

8. Prove the following equations:

$$\vec{E}_{1//} = \vec{E}_{2//}$$

$$\vec{D}_{1\perp} - \vec{D}_{2\perp} = \rho \qquad \vec{H}_{1//} - \vec{H}_{2//} = \vec{J}$$

$$\vec{B}_{1\perp} = \vec{B}_{2\perp}$$

9. Discuss a relation between the electric displacement entering a medium if there is no surface charge.

10. Discuss a relation between the magnetic field strength entering a medium if there is no surface current.

Chapter 6

Methods of Determining Electric Field and Potential

6.1 Introduction

In Chapters 3 and 4, we learned the simplest way to calculate electric field emerging from a single charge. In this chapter, we will study more general ways to calculate electric field emerging from charge distributions in conductors and dielectrics. We will learn how to solve the Laplace–Poisson equation under electromagnetic boundary conditions, as described in Chapter 5. A solution to the Laplace–Poisson equation requires mathematical skills in some cases, but in other cases requires only a very intuitive method, called the method of images. More generally, the Laplace–Poisson equation can be solved by using Green's function.

6.2 Laplace–Poisson Equation

From Maxwell's equations (Eqs. 1.26 and 2.1) in Chapters 1 and 2 and the scalar potential (Eq. 3.12 in which we use the symbol ϕ instead of V if the absolute potential at infinity becomes 0) defined in Chapter 3, we have

$$\vec{\nabla} \cdot \vec{D} = \rho_{\text{TRUE}}$$

$$\vec{D} = \varepsilon \vec{E}$$

$$\vec{E} = -\vec{\nabla} \phi$$

Optical Properties of Solids: An Introductory Textbook
Kitsakorn Locharoenrat
Copyright © 2016 Pan Stanford Publishing Pte. Ltd.
ISBN 978-981-4669-06-1 (Hardcover), 978-981-4669-07-8 (eBook)
www.panstanford.com

By combining these equations, we obtain in Cartesian coordinates:

$$\Delta\phi = \frac{\partial^2\phi}{\partial x^2} + \frac{\partial^2\phi}{\partial y^2} + \frac{\partial^2\phi}{\partial z^2} = -\frac{\rho_{\text{TRUE}}}{\varepsilon} \tag{6.1}$$

Here we assume that ε does not vary as a function of time. Equation 6.1 is known as the *Laplace–Poisson* equation. When ρ_{TRUE} is equal to zero, it is called the Laplace equation, and when it is not equal to zero, it is called the Poisson equation. In vacuum, Eq. 6.1 is modified as

$$\Delta\phi = \frac{\partial^2\phi}{\partial x^2} + \frac{\partial^2\phi}{\partial y^2} + \frac{\partial^2\phi}{\partial z^2} = -\frac{\rho_{\text{TOTAL}}}{\varepsilon_0} \tag{6.2}$$

For convenience, we rewrite Eq. 6.1 in polar coordinate (r, θ, φ) and cylindrical coordinate (r, φ, z) as follows (Fig. 6.1):

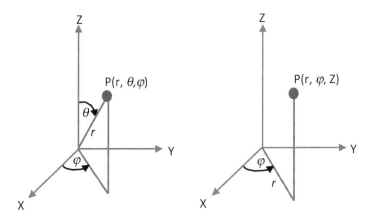

Figure 6.1 Polar coordinates (left) and cylindrical coordinates (right) of the observation point *P*.

In spherical coordinates, the equation is

$$\Delta\phi = \frac{1}{r^2}\left\{\frac{\partial}{\partial r}\left(r^2\frac{\partial\phi}{\partial r}\right) + \frac{1}{\sin\theta}\frac{\partial}{\partial\theta}\left(\sin\theta\frac{\partial\phi}{\partial\theta}\right) + \frac{1}{\sin^2\theta}\frac{\partial^2\phi}{\partial\varphi^2}\right\} = -\frac{\rho_{\text{TRUE}}}{\varepsilon} \tag{6.3}$$

In cylindrical coordinates, the equation is

$$\Delta\phi = \frac{1}{r}\frac{\partial}{\partial r}\left(r\frac{\partial\phi}{\partial r}\right) + \frac{1}{r^2}\frac{\partial^2\phi}{\partial\varphi^2} + \frac{\partial^2\phi}{\partial z^2} = -\frac{\rho_{\text{TRUE}}}{\varepsilon} \tag{6.4}$$

When a distribution of charges is given, the electric potential can be found everywhere by using this equation. At boundaries between dielectrics, the following boundary conditions hold for scalar potential:

$$\phi_1 = \phi_2 \tag{6.5}$$

For metals, $\vec{E} \perp$ surfaces

$$\varepsilon_1 \frac{\partial \phi_1}{\partial \xi} = \varepsilon_2 \frac{\partial \phi_2}{\partial \xi} \tag{6.6}$$

This means that the electric potential ϕ is determined by Eqs. 6.1 and 6.6 uniquely. Namely, the specification of a potential under the closed surface determines the outstanding potential distribution in a space surrounded by the surface (Dirichlet boundary conditions). The specification of an electric field everywhere under the closed surface also determines a unique potential distribution (Neumann boundary conditions).

There are four ways of solving the Laplace–Poisson equation under given electromagnetic boundary conditions: method of images, a direct solution, conformal mapping, and Green's function. All solutions will be explained in the following sections.

6.3 Method of Images

The method of images consists of the following five steps. From Fig. 6.2, we first assume the existence of an imaginary charge, instead of directly solving the Laplace–Poisson equation. This imaginary charge is known as an image charge.

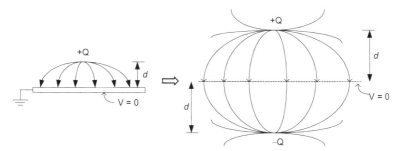

Figure 6.2   A positive charge lies a distance *d* above the conducting surface images.

1. We consider an area with a uniform dielectric constant and assume a charge distribution ρ_1. We also assume electromagnetic boundary conditions at the boundary of this area.

2. If we consider the electric potential ϕ_1 due to a charge distribution ρ_1 alone, it will fulfill the Laplace–Poisson equation, but it will not fulfill the boundary conditions in general.

3. Then we put the image charges outside this area and superpose the electric potential ϕ_2 onto ϕ_1. We assume that the same dielectric fills an outside space. We see that $\Delta\phi_2 = 0$ inside the area because ϕ_2 is the electric potential due to the charge outside the area. Hence, we have $\Delta(\phi_1 + \phi_2) = \rho_1$ inside the area.

4. The electric potential $\phi_1 + \phi_2$ at the boundary differs from ϕ_1. If ϕ_2 or image charges are appropriately chosen, the boundary condition will be fulfilled. Therefore, the method of images can be regarded as a maneuvering of the boundary values of the electric potential by the 'image charges' *outside* the area in problem.

5. From the theory explaining the operation of electric shield, the electric potential found by steps (1) to (4) is a unique solution to the Laplace–Poisson equation.

On the other hand, there are three boundary conditions used in this method:

1. The surface of the conductor is suggested to be the equipotential surface ($\phi = 0$, $E_x = E_y = 0$).

2. The electric displacement $\vec{D}$ is continuous across a boundary. Namely, the normal component of the electric displacement does not vary from one side of the boundary to the other.

3. The tangential component of electric field is the same on both sides of the boundary.

This calculation method starts from Gauss's law:

$$\int_S \vec{E} \cdot \vec{n}\, dS = \frac{1}{\varepsilon}Q \tag{6.7}$$

and we get

$$(E_{1\perp} - E_{2\perp})A = \frac{Q}{\varepsilon_0} = \frac{\sigma A}{\varepsilon_0} \tag{6.8}$$

$$E_{1\perp} = \frac{\sigma}{\varepsilon_0} \tag{6.9}$$

From the electric potential due to the charges,

$$\phi = \frac{1}{4\pi\varepsilon} \sum_i \frac{q_i}{r_i} \tag{6.10}$$

we also get

$$\phi_1(r,\phi,z) = \frac{1}{4\pi\varepsilon_0}\left(\frac{+q}{\sqrt{r^2+(z-d)^2}} + \frac{-q}{\sqrt{r^2+(z+d)^2}} \right) \tag{6.11}$$

From Eq. 6.9 together with the boundary condition of scalar potential, we finally get

$$\sigma(r,\varphi,0) = \varepsilon_0 E_{1\perp}(r,\varphi,0) = -\varepsilon_0 \frac{\partial \phi_1}{\partial z}\Big|_{z=0} = \frac{-1}{4\pi}\frac{2qd}{(r^2+d^2)^{1.5}} \tag{6.12}$$

$$Q = -\frac{1}{4\pi}\int_0^\infty\int_0^{2\pi} \frac{2qd}{(r^2+d^2)^{1.5}} r\,d\varphi\,dr = -q \tag{6.13}$$

6.4 Direct Solution

The Laplace equation in one dimension can be easily solved. When the boundary conditions have spherical or cylindrical symmetry, we can solve Eqs. 6.3 and 6.4 directly. We often solve the Laplace equation ($\rho_{\text{TRUE}} = 0$) where there is no charge, no conductor, and no dielectrics. The solution to the Laplace equation in a space with spherical or azimuthal symmetry will finally be as follows:

In spherical coordinates:

$$\phi = \sum_{n,m}\left(a_n r^n + \frac{b_n}{r^{n+1}} \right)(c_m \cos m\varphi + d_m \sin m\varphi)(P_n^m \cos\theta) \tag{6.14}$$

In cylindrical coordinates:

$$\phi = \sum_{n=1}^\infty (a_n \cos n\varphi + b_n \sin n\varphi)\left(c_n r^n + \frac{d_n}{r^n} \right) + (a_0\varphi + b_0)(c_0 \log r + d_0) \tag{6.15}$$

Here, $P_n^m \cos\theta$ is the associated Legendre function.

The coefficients (a_n, b_n, c_n, etc.) are determined so that the electromagnetic boundary conditions are fulfilled. It is noted that even if electric fields due to charges outside the area enter the area with $\rho = 0$, the Laplace equation $\Delta\phi = 0$ still holds.

6.5   Confocal Mapping

Electric fields in two dimensions can be found by solving the Laplace equation by using the property of complex function $W(Z)$. This method was well-known in earlier days. Both the real component U and the imaginary component V of a holomorphic function of complex variables,

$$W(Z) = U(X + iY) + iY) + iV(X + iY) \tag{6.16}$$

for which dW/dZ exists, are solutions to the Laplace equation. The curves $U(X + iY) = $ constant and $V(X + iY) = $ constant are orthogonal to each other. When one of two curves represents the equipotential curve, the other shows the lines of electric forces. Then, if we consider a metal object with a shape $U(X + iY) = $ constant, the equipotential curve around it is $U(X + iY)$ and the line of electric force is $V(X + iY) = $ constant. The inconvenience in using this method is that we have to know a priori the complex function $W(Z)$ representing the shape of the conductor object.

6.6   Green's Function

$G(x, y, z | \xi, \eta, \zeta)$ is the electric potential at any point (x, y, z) when a unit charge is put at (ξ, η, ζ) under a given boundary condition. The electric potential due to the charge distribution $\rho(\xi, \eta, \zeta)$ is equal to the integration of $G(x, y, z | \xi, \eta, \zeta)$ multiplied by $\rho(\xi, \eta, \zeta)$. The electromagnetic Green's function in free space will finally be

$$G(x,y,z|\xi,\mu,\zeta) = \frac{1}{4\pi} \cdot \frac{1}{\sqrt{(x-\xi)^2 + (y-\eta)^2 + (z-\zeta)^2}} \tag{6.17}$$

Green's function is a very useful tool in computational calculation. It is utilized in the complex calculation of electric potential.

6.7 Problems

1. Prove that

$$\Delta\phi = \frac{1}{r^2}\left\{\frac{\partial}{\partial r}\left(r^2\frac{\partial\phi}{\partial r}\right) + \frac{1}{\sin\theta}\frac{\partial}{\partial\theta}\left(\sin\theta\frac{\partial\phi}{\partial\theta}\right) + \frac{1}{\sin^2\theta}\frac{\partial^2\phi}{\partial\varphi^2}\right\} = -\frac{\rho_{TRUE}}{\varepsilon}$$

2. Prove that

$$\Delta\phi = \frac{1}{r}\frac{\partial}{\partial r}\left(r\frac{\partial\phi}{\partial r}\right) + \frac{1}{r^2}\frac{\partial^2\phi}{\partial\varphi^2} + \frac{\partial^2\phi}{\partial z^2} = -\frac{\rho_{TRUE}}{\varepsilon}$$

3. Discuss the Dirichlet boundary conditions and Neumann boundary conditions.

4. Find an image charge for a metal plane and a point charge in front of it.

5. Find image charges of a grounded metal sphere with a nearby point charge.

6. Find the electric potential due to a uniform charge distribution in a layer.

7. Prove that the electric potential in spherical coordinates is

$$\phi = \sum_{n,m}\left(a_n r^n + \frac{b_n}{r^{n+1}}\right)(c_m\cos m\varphi + d_m\sin m\varphi)(P_n^m\cos\theta)$$

8. Prove that the electric potential in cylindrical coordinates is

$$\phi = \sum_{n=1}^{\infty}(a_n\cos n\varphi + b_n\sin n\varphi)\left(c_n r^n + \frac{d_n}{r^n}\right) + (a_0\varphi + b_0)(c_0\log r + d_0)$$

9. Plot curves U = constant and V = constant when $W = \cosh^{-1}(aZ)$.

10. Prove that the electric potential using Green's function is

$$G(x,y,z|\xi,\mu,\zeta) = \frac{1}{4\pi}\cdot\frac{1}{\sqrt{(x-\xi)^2 + (y-\eta)^2 + (z-\zeta)^2}}$$

Chapter 7

Light

7.1 Introduction

We will make a brief survey of the basic concepts of light fields. Then we will move on to de Broglie waves, cable waves, and electromagnetic waves.

7.2 Light

In quantum theory, light is simultaneously a wave and a particle according to de Broglie's famous formula.

Light is composed of *electromagnetic waves* in which the electric and magnetic field amplitudes are replaced by operators. The electromagnetic spectrum represents the wavelength region of the light, as shown in Table 7.1.

Light is also composed of particles called "photons" with their photon energy given by

$$h\upsilon = h[\text{Js}]\frac{c[\text{m/s}]}{\lambda[\text{m}]} \tag{7.1}$$

$$h\upsilon = \frac{h[\text{Js}]\,c[\text{m/s}]\,e[\text{eV/J}]}{\lambda[\text{m}]} \tag{7.2}$$

Optical Properties of Solids: An Introductory Textbook
Kitsakorn Locharoenrat
Copyright © 2016 Pan Stanford Publishing Pte. Ltd.
ISBN 978-981-4669-06-1 (Hardcover), 978-981-4669-07-8 (eBook)
www.panstanford.com

$$hv = \frac{h[Js]\,c[10^9\,nm/s]\,e[eV/J]}{\lambda[nm]} \qquad (7.3)$$

By substituting Plank's constant ($h = 6.626 \times 10^{-34}$ J.s), light speed ($c = 2.998 \times 10^8$ m/s) and energy ($e = 6.242 \times 10^{18}$ eV), we will get

$$hv = \frac{1240}{\lambda[nm]}\ [eV] \qquad (7.4)$$

1 eV = 8066 cm^{-1} wave number (cm^{-1})
1 Ry = 13.6 eV   Rydberg (Ry)   [RY] = [AU]
1 AU (1 Hartree) = 27.2 eV
The speed of light is $c = 2.998 \times 10^8$ m/s

Table 7.1   Electromagnetic spectrum

Light	Wavelength	Details
Infrared (IR)	800 nm to 100 μm	≤ 2.5 μm: near IR
		≥ 40 μm: far IR
Visible (VIS)	400 nm to 800 nm	Red
		Orange
		Yellow
		Green
		Blue
		Indigo
		Violet
Ultraviolet (UV)	100 nm to 400 nm	≤ 200 nm: vacuum UV

7.3 De Broglie Waves

In 1929 de Broglie received the Nobel Prize in physics for discovering the wave nature of electrons. "Electrons, like photons (particles of light waves), can act like a particle and a wave," said de Broglie. With this discovery, he introduced a new field of study in the science of quantum physics or wave mechanics.

A basic mechanical wave is shown in Fig. 7.1. Mechanical waves are disturbances through a medium (i.e., air, water, or vacuum) that usually transfer energy. They are described using several terms such as wavelength, period, frequency, amplitude, and speed of propagation.

The distance between each bump is called wavelength (λ). The number of bumps per second is called frequency ($f = 1/T$). T is time period in sec. In general, the velocity at which a wave moves is proportional to λ and f and is given by

$$v = f\lambda \tag{7.5}$$

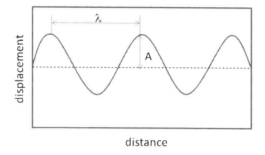

distance

Figure 7.1 A mechanical wave.

Herein, two velocities are associated with waves: phase and group velocity. Phase velocity (v_p) is the velocity of oscillation (phase) of waves. It is proportional to ω and inversely proportional to k (for details, see Appendix A) or

$$v_p = \frac{\omega}{k} \tag{7.6}$$

For any electromagnetic wave, phase velocity is equal to the speed of light ($v_p = c$).

Group velocity (v_g) is the velocity at which the energy of waves propagates. It is equal to the derivative of ω with respect to k or

$$v_g = \frac{d\omega}{dk} \tag{7.7}$$

Since group velocity is the derivative of phase velocity, phase velocity will often be greater than group velocity. When dealing with waves going in oscillations (cycles of periodic movements), we use the notations of angular frequency (ω) and wave number (k), which is inversely proportional to λ. The equations for both are as follows:

$$\omega = 2\pi f \tag{7.8}$$

$$k = \frac{2\pi}{\lambda} \tag{7.9}$$

The relationship between photoelectric and Compton effects is shown in Fig. 7.2.

Figure 7.2   Photoelectric (left) and Compton (right) effects.

$$E = hf \text{ (Planck's formula)} \tag{7.10}$$

$$E = mc^2 \text{ (Einstein's formula)} \tag{7.11}$$

Let us join the two equations as

$$E = hf = mc^2 \tag{7.12}$$

where the Planck's constant h is equal to 6.63×10^{-34} Js. The energy E is 1 eV (= 1.6×10^{-19} J), and the speed of light (c) is equal to 3×10^8 m/s. If we define the momentum of photon as

$$p = mc \tag{7.13}$$

and we substitute Eq. 7.13 in Eq. 7.12, we get

$$hf = pc \tag{7.14}$$

$$\frac{h}{p} = \frac{c}{f} \tag{7.15}$$

Substituting what we know for wavelength ($\lambda = \dfrac{v}{f}$ or in this case $\dfrac{c}{f}$), we get

$$\frac{h}{p} = \lambda \tag{7.16}$$

This is the equation de Broglie discovered in his 1924 doctoral thesis. It accounts for both waves and particles, mentioning momentum (particle aspect) and wavelength (wave aspect). Explaining his hypothesis, he associated two wave velocities with particles (not just photons). Now considering the phase velocity of particle with momentum $p = mv$ and substituting f from Eq. 7.8 in Eq. 7.10, we get

$$\omega = 2\pi f = \frac{2\pi E}{h} \tag{7.17}$$

Also substituting λ from Eq. 7.16 in Eq. 7.9, we get

$$k = \frac{2\pi}{\lambda} = \frac{2\pi\, p}{h} \tag{7.18}$$

Equating these two equations, we get

$$v_{\mathrm{p}} = \frac{\omega}{k} = \frac{E}{p} \tag{7.19}$$

From this new equation of phase velocity of particle, we can derive

$$v_{\mathrm{p}} = \frac{E}{p} \tag{7.20}$$

$$v_{\mathrm{p}} = \frac{mc^2}{mv} \tag{7.21}$$

$$v_{\mathrm{p}} = \frac{c^2}{v} \tag{7.22}$$

It has been shown that the actual velocity of particle (v) is smaller than the velocity of light (c). If the relativistic theory for particles is defined as

$$E = \sqrt{p^2 c^2 + m^2 c^4} \tag{7.23}$$

and we substitute Eq. 7.23 in Eq. 7.7, we get

$$v_{\mathrm{g}} = \frac{d\omega}{dk} = \frac{dE}{dp} \tag{7.24}$$

$$v_{\mathrm{g}} = \frac{d\sqrt{p^2 c^2 + m^2 c^4}}{dp} \tag{7.25}$$

$$v_{\mathrm{g}} = \frac{pc^2}{E} = \frac{mvc^2}{mc^2} \tag{7.26}$$

$$v_{\mathrm{g}} = v \tag{7.27}$$

So the group velocity of a particle is equal to the actual velocity of the particle.

If a particle has a wave property ($v \to c$), Eq. 7.22, we finally get

$$v_{\mathrm{p}} = c \tag{7.28}$$

Not only is de Broglie's equation useful for small particles (i.e., electrons, protons), but also can be applied to larger particles such as everyday objects. Essentially, if we imagine a particle (or a miniature man) traveling on a phase of wave, we can measure these conditions under the particle's velocity.

7.4 Cable Waves

In this section, we will investigate the wave propagation of voltages and currents. Voltage and current waves are usually sent along a geometrical configuration of wires and cables known as transmission lines, as shown in Fig. 7.3. A transmission line is a medium for propagating energy from one point to another. Propagation of energy is required for one of the two general reasons: power transfer (i.e., for lighting and heating) and information transfer (i.e., for telephone and radio).

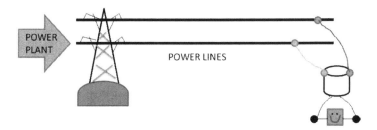

Figure 7.3   Power line system.

Transmission lines are represented by a pair of parallel wires; a generator feeds power into one end of these lines. In the absence of any resistance in the lines, the cable is known as ideal or lossless. In contrast, if any resistance (i.e., resistor, conductor) exists in the line, the cable is known as real or lossy.

Before moving on to transmission lines, we will first understand the basic concept of wave equation (for more details, see Appendix A). If waves travel in the x-direction (Fig. 7.4) with velocity v, a general solution to the wave equation is

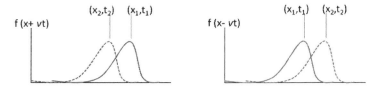

Figure 7.4   Backward- (left) and forward-moving waves.

$$A = f(x \pm vt) \tag{7.29}$$

Taking a space derivative, we get

$$\frac{\partial A}{\partial x} = f'(x \pm vt) \tag{7.30}$$

$$\frac{\partial^2 A}{\partial x^2} = f''(x \pm vt) \tag{7.31}$$

Taking a time derivative, we obtain

$$\frac{\partial A}{\partial t} = \pm v\, f'(x \pm vt) \tag{7.32}$$

$$\frac{\partial^2 A}{\partial t^2} = v^2\, f''(x \pm vt) \tag{7.33}$$

Dividing Eq. 7.31 by Eq. 7.33, we get

$$\frac{\partial^2 A}{\partial x^2} = \frac{1}{v^2} \frac{\partial^2 A}{\partial t^2} \tag{7.34}$$

If the Laplacian of A is defined as

$$\nabla^2 A = \left(\frac{\partial^2}{\partial x^2} + \frac{\partial^2}{\partial y^2} + \frac{\partial^2}{\partial z^2} \right) A \tag{7.35}$$

then the generalized form of a wave equation will be

$$\nabla^2 A = \frac{1}{v^2} \frac{\partial^2 A}{\partial t^2} \tag{7.36}$$

Here A could be either voltage (V) or current (I) as in waves in transmission lines or electric field ($\vec{E}$) or magnetic field ($\vec{H}$) as in electromagnetic waves in free space.

Let us suppose lossless power lines as a chain of very short cables (Fig. 7.5), including inductor L (unit: H/m or Ωs/m) and capacitor C (unit: F/m or s/Ωm), each carrying a circulating current and inducing voltage.

According to Kirchhoff's voltage law for summing voltages, we have

$$V = V_C + V_L \tag{7.37}$$

$$V = V + \left(\frac{\partial V}{\partial x} \right) dx + L \left(\frac{\partial I}{\partial t} \right) dx \tag{7.38}$$

$$\frac{\partial V}{\partial x} = -L \frac{\partial I}{\partial t} \tag{7.39}$$

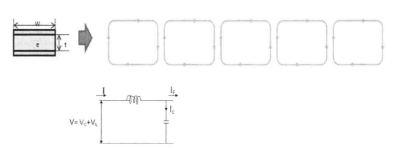

Figure 7.5 Lossless power lines.

Similarly, using Kirchhoff's current law for summing currents, we have

$$I = I_C + I_F \tag{7.40}$$

If we define $I_C = C\left(\dfrac{\partial V}{\partial t}\right)dx$ and $I_F = I + \left(\dfrac{\partial I}{\partial x}\right)dx$, we will get

$$I = C\left(\frac{\partial V}{\partial t}\right)dx + I + \left(\frac{\partial I}{\partial x}\right)dx \tag{7.41}$$

$$\frac{\partial I}{\partial x} = -C \frac{\partial V}{\partial t} \tag{7.42}$$

Equations 7.39 and 7.42 describe the propagation of guided voltage and current waves on lossless transmission lines (also called *Telegrapher's equations*).

Differentiating Eqs. 7.39 and 7.42 with respect to space, we get

$$\frac{\partial^2 V}{\partial x^2} = -L \frac{\partial}{\partial t}\frac{\partial I}{\partial x} \tag{7.43}$$

$$\frac{\partial^2 I}{\partial x^2} = -C \frac{\partial}{\partial t}\frac{\partial V}{\partial x} \tag{7.44}$$

Substituting Eq. 7.42 in Eq. 7.43 and Eq. 7.39 in Eq. 7.44, we have

$$\frac{\partial^2 V}{\partial x^2} = LC \frac{\partial^2 V}{\partial t^2} \tag{7.45}$$

$$\frac{\partial^2 I}{\partial x^2} = LC \frac{\partial^2 I}{\partial t^2} \tag{7.46}$$

Comparing with Eq. 7.34, we get

$$v = \sqrt{\frac{1}{LC}} \qquad (7.47)$$

From Eqs. 7.45 and 7.46, we obtain a solution to voltage and current waves moving forward (subscript F) and backward (subscript B) through transmission lines as

$$V(x, t) = V_F e^{-i(kx - \omega t)} + V_B e^{i(kx + \omega t)} \qquad (7.48)$$

$$I(x, t) = I_F e^{-i(kx - \omega t)} + I_B e^{i(kx + \omega t)} \qquad (7.49)$$

If we try a solution for V in Eq. 7.48 of the simplified form as

$$V = V_B e^{i(kx + \omega t)} \qquad (7.50)$$

and substitute Eq. 7.50 in Eq. 7.45, we get

$$\frac{\partial^2 V}{\partial x^2} = LC \frac{\partial^2 V}{\partial t^2} \qquad (7.51)$$

$$(ik)^2 = LC(i\omega)^2 \qquad (7.52)$$

Then, the phase constant (Fig. 7.6) will be

$$k = \omega\sqrt{LC} \qquad (7.53)$$

$$k = \frac{\omega}{v} \qquad (7.54)$$

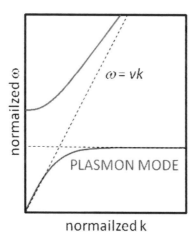

Figure 7.6 Angular frequency versus phase constant.

In the case of real power lines as shown in Fig. 7.7, we will include a resistor R (unit: Ω/m) along the line and a conductor G (unit: S/m) across the line, which will be responsible for energy losses.

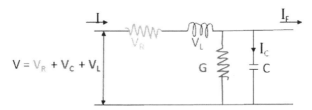

Figure 7.7 Lossy power lines.

Using Kirchhoff's voltage law to sum voltages, we have

$$V = V_R + V_C + V_L \tag{7.55}$$

If we define, $V_R = RIdx$, $V_C = V + \left(\dfrac{\partial V}{\partial x}\right)dx$ and $V_L = L\left(\dfrac{\partial I}{\partial t}\right)dx$, we will get

$$V = RIdx + V + \left(\frac{\partial V}{\partial x}\right)dx + L\left(\frac{\partial I}{\partial t}\right)dx \tag{7.56}$$

$$\frac{\partial V}{\partial x} = -\left[RI + L\left(\frac{\partial I}{\partial t}\right)\right] \tag{7.57}$$

Similarly, using Kirchhoff's current law to sum currents, we get

$$I = I_G + I_C + I_F \tag{7.58}$$

If we define, $I_G = GVdx$, $I_C = C\left(\dfrac{\partial V}{\partial t}\right)dx$ and $I_F = I + \left(\dfrac{\partial I}{\partial x}\right)dx$, we will get

$$I = GVdx + C\left(\frac{\partial V}{\partial t}\right)dx + I + \left(\frac{\partial I}{\partial x}\right)dx \tag{7.59}$$

$$\frac{\partial I}{\partial x} = -\left[GV + C\left(\frac{\partial V}{\partial t}\right)\right] \tag{7.60}$$

Considering a simple solution for V and I, we get

$$V = V_B e^{i(k'x + \omega t)} \tag{7.61}$$

$$I = I_B e^{i(k'x + \omega t)} \tag{7.62}$$

Substituting Eq. 7.62 in Eq. 7.57, we get

$$\frac{\partial V}{\partial x} = -\left[RI + L\left(\frac{\partial I}{\partial t}\right)\right] \text{ (which is Eq. 7.57)]}$$

$$\frac{\partial V}{\partial x} = -\left[RI + LI_B\left(\frac{\partial}{\partial t}e^{i(kx+\omega t)}\right)\right] \tag{7.63}$$

$$\frac{\partial V}{\partial x} = -\left[RI + i\omega LI_B e^{i(kx+\omega t)}\right] \tag{7.64}$$

$$\frac{\partial V}{\partial x} = -(R + i\omega L)I \tag{7.65}$$

Substituting Eq. 7.61 in Eq. 7.60, we get

$$[\frac{\partial I}{\partial x} = -GV + C\left(\frac{\partial V}{\partial t}\right) \text{ (which is Eq. 7.60)]}$$

$$\frac{\partial I}{\partial x} = -\left[GV + CV_B\left(\frac{\partial}{\partial t}e^{i(kx+\omega t)}\right)\right] \tag{7.66}$$

$$\frac{\partial I}{\partial x} = -\left[GV + i\omega CV_B e^{i(kx+\omega t)}\right] \tag{7.67}$$

$$\frac{\partial I}{\partial x} = -(G + i\omega C)V \tag{7.68}$$

If we can write the expression for a lossy line starting from that of a lossless line ($L \to L'$ and $C \to C'$) by defining

$$\frac{\partial V}{\partial x} = -L'\frac{\partial I}{\partial t} = -i\omega\, L'\, I \text{ (which is Eq. 7.39)}$$

$$\frac{\partial I}{\partial x} = -C'\frac{\partial V}{\partial t} = -i\omega\, C'\, V \text{ (which is Eq. 7.42)}$$

we get

$$i\omega\, L' = R + i\omega\, L \tag{7.69}$$

$$i\omega\, C' = G + i\omega\, C \tag{7.70}$$

Also $k = \omega\sqrt{LC}$ in a lossless line corresponds to

$$k' = \omega\sqrt{L'C'} \tag{7.71}$$

Substituting Eqs. 7.69 and 7.70 in Eq. 7.71, we get

$$k' = \frac{\omega}{i\omega}\sqrt{(R + i\omega L)(G + i\omega C)} \tag{7.72}$$

Defining a real term α corresponding to attenuation along the line, which is known as the *attenuation* or *absorption coefficient*, we get

$$\alpha + ik = ik' = \sqrt{(R + i\omega L)(G + i\omega C)} \tag{7.73}$$

Substituting *ik'* with $\alpha + ik$ in Eqs. 7.48 and 7.49, we get

$V(x, t) = V_F e^{-i(k'x - \omega t)} + V_B e^{i(k'x + \omega t)}$ which is Eq. 7.48 and we use *k'* (lossy lines) instead of *k* (lossless lines).

$I(x, t) = I_F e^{-i(k'x - \omega t)} + I_B e^{i(k'x + \omega t)}$ which is Eq. 7.49 and we use k' (lossy lines) instead of k (lossless lines).

$$V(x, t) = V_F e^{-ax - i(kx - \omega t)} + V_B e^{ax + i(kx + \omega t)} \tag{7.74}$$

$$I(x, t) = I_F e^{-ax - i(kx - \omega t)} + I_B e^{ax + i(kx + \omega t)} \tag{7.75}$$

For a forward voltage wave (Fig. 7.8), we have

$$V = V_F e^{-ax + i(kx + \omega t)} \tag{7.76}$$

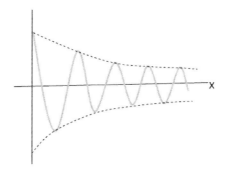

Figure 7.8 Forward voltage waves in lossy lines.

Recalling the telegrapher's equation as

$$\frac{\partial I}{\partial x} = -C \frac{\partial V}{\partial t} \tag{7.77}$$

Substituting the following solutions in Eq. 7.77

$$V(x, t) = V_F e^{-i(kx - \omega t)} + V_B e^{i(kx + \omega t)} \qquad \text{(which is Eq. 7.48)}$$

$$I(x, t) = I_F e^{-i(kx - \omega t)} + I_B e^{i(kx + \omega t)} \qquad \text{(which is Eq. 7.49)}$$

we get

$$-ikI_F e^{-i(kx - \omega t)} + ikI_B e^{i(kx + \omega t)} = -C \left[i\omega V_F e^{-i(kx - \omega t)} + i\omega V_B e^{i(kx + \omega t)} \right] \tag{7.78}$$

Since $e^{i(kx - \omega t)}$ and $e^{i(kx + \omega t)}$ represent waves traveling in opposite directions, they can be treated separately. This leads to two independent expressions in I and V:

$$-i\omega C V_F e^{-i(kx - \omega t)} = -ikI_F e^{i(kx - \omega t)} \tag{7.79}$$

$$\frac{V_F}{I_F} = \frac{k}{\omega C} \tag{7.80}$$

$$-i\omega C V_B e^{-i(kx - \omega t)} = -ikI_B e^{i(kx - \omega t)} \tag{7.81}$$

$$\frac{V_B}{I_B} = \frac{-k}{\omega C} \tag{7.82}$$

Substituting $k = \omega\sqrt{LC}$ in Eq. 7.80, we get

$$\frac{V_F}{I_F} = \frac{\omega\sqrt{LC}}{\omega C} = \sqrt{\frac{L}{C}} \tag{7.83}$$

If the *characteristic impedance* Z_0 is defined as the ratio of the V and I of a unidirectional (unreflected) wave in a transmission line at any point without reflection, we get (in lossless lines)

$$Z_0 = \frac{V_F}{I_F} = -\frac{V_B}{I_B} = \sqrt{\frac{L}{C}} \quad \text{(unit: } \Omega) \tag{7.84}$$

In lossy lines, we get

$$Z_0' = \sqrt{\frac{L'}{C'}} \tag{7.85}$$

Substituting Eqs. 7.69 and 7.70 in Eq. 7.85, we get

$$Z_0' = \sqrt{\frac{(R + i\omega L)}{(G + i\omega C)}} \tag{7.86}$$

In principle, when a transmission line has waves only in the positive direction, V and I waves are always in phase (Fig. 7.9), energy is propagated, and a generator feeds power into the line at all times.

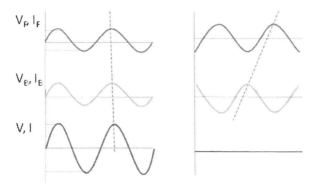

Figure 7.9   Waves in same (left) and opposite (right) directions.

In practice, this situation is destroyed when waves travel in both directions; waves in the negative direction are produced by reflection at the boundary when a line is terminated or mismatched.

Consider a load added to the end of a transmission line as shown in Fig. 7.10.

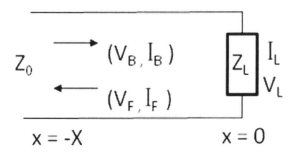

Figure 7.10 Load at the power lines.

And also define $V(x, t) = V(x)V(t)$ and $I(x, t) = I(x)I(t)$

$V(x, t) = V_F e^{-i(kx - \omega t)} + V_B e^{i(kx + \omega t)}$ (which is Eq. 7.48)

$I(x, t) = I_F e^{-i(kx - \omega t)} + I_B e^{-i(kx + \omega t)}$ (which is Eq. 7.49)

$$V(x)V(t) = (V_F e^{-ikx} + V_B e^{ikx})e^{i\omega t} \tag{7.87}$$

$$I(x)I(t) = (I_F e^{-ikx} + I_B e^{ikx})e^{i\omega t} \tag{7.88}$$

At the load L $(x = 0)$, we get

$$I(0) = I_F + I_B = I_L \tag{7.89}$$

$$V(0) = V_F + V_B = V_L = Z_L I_L = Z_L(I_F + I_B) \tag{7.90}$$

$$V_F + V_B = \frac{Z_L(V_F - V_B)}{Z_0} \tag{7.91}$$

Reflection coefficient (ρ) is defined as the amplitude of the reverse voltage wave divided by the amplitude of the forward voltage wave at the load and is given by

$$\rho = \frac{V_B}{V_F} = \frac{Z_L - Z_0}{Z_L + Z_0} \tag{7.92}$$

Similar to V, we get

$$\rho = \frac{I_B}{I_F} = \frac{Z_0 - Z_L}{Z_L + Z_0} \tag{7.93}$$

We can see that the reflection coefficient can have any value in the range $-1 \le \rho \le 1$.

Reflected power from the load is defined as

$$R = \rho^2 = \left(\frac{Z_L - Z_0}{Z_L + Z_0}\right)^2 \tag{7.94}$$

This reflection results in the setting up of standing waves in the transmission line. On the other hand, the voltage standing wave ratio (VSWR) is the ratio of maximum V to minimum V and it can be stated in terms of the reflection coefficient by

$$\text{VSWR} = \frac{V_F + V_B}{V_F - V_B} = \frac{1+\rho}{1-\rho} \tag{7.95}$$

or the reflection coefficient can be stated in terms of VSWR by

$$\rho = \frac{\text{VSWR} - 1}{\text{VSWR} + 1} \tag{7.96}$$

Equation 7.96 shows that a zero reflection requires $\rho = 0$ leading to VSWR = 1 (i.e., short circuit end or $Z_L = Z_0$). In this case, we get all the power to the load and there are no echoes (*matching*).

By contrast, a multiple reflection requires $\rho = 1$ leading to VSWR = ∞ (i.e., open circuit end or $Z_L = \infty$). This also causes a phenomena called *ringing*. Ringing is the unwanted oscillations of voltage and/or current caused by multiple reflections. For instance, the original wave is reflected at the load; this reflection then gets reflected back at the generator (*G*). We will show this effect by looking at a step change in voltage when a device is switched on and a pulse V_F is generated, which travels toward the load, as shown in Fig. 7.11.

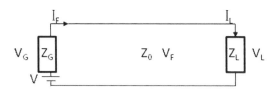

Figure 7.11   Power line systems.

Using Kirchhoff's voltage law to sum voltages, we get

$$V_P^1 = V_F = V - V_G = V - (I_F Z_G) = V - \frac{V_F Z_G}{Z_0} \tag{7.97}$$

Part of the pulse V_P^1 is then reflected at the load as

$$V_P^2 = V_B = r_L V_F \tag{7.98}$$

Next V_P^2 is reflected at generator G as

$$V_P^3 = r_G V_2 = r_L r_G V_F \qquad (7.99)$$

The asymptotic value is

$$V_P^1 = V_F$$

$$V_P^2 = V_F + \rho_L V_F = V_F(1 + \rho_L)$$

$$V_P^3 = V_F + \rho_L V_F + \rho_L \rho_G V_F = V_F(1 + \rho_L + \rho_L \rho_G)$$

$$V_P^4 = V_F + \rho_L V_F + \rho_L \rho_G V_F + \rho_L^2 \rho_G V_F = V_F(1 + \rho_L + \rho_L \rho_G + \rho_L^2 \rho_G)$$

$$V_P^n = V_F(1 + \rho_L + \rho_L \rho_G + \rho_L^2 \rho_G + \rho_L^2 \rho_G^2 + \rho_L^3 \rho_G^2 + ...)$$

$$V_P^n = V_F(1 + \rho_L \rho_G + \rho_L^2 \rho_G^2 + ...) + V_F \rho_L(1 + \rho_L \rho_G + \rho_L^2 \rho_G^2 + ...)$$

$$V_P^\infty = V_F + V_F r_L \sum (\rho_L \rho_G)^n \qquad (7.100)$$

By definition of $\rho_{G,R} \leq 1$ (or $x \leq 1$ in our case), we have

$$\sum x^n = \frac{1}{1-x} \qquad (7.101)$$

Then

$$V_P^\infty = \frac{V_F(1 + \rho_L)}{1 - \rho_L \rho_G} \qquad (7.102)$$

From Eq. 7.97, we get

$$V_F = \frac{Z_0 V}{Z_0 + Z_G} \qquad (7.103)$$

and define

$$\rho_L = \frac{Z_L - Z_0}{Z_L + Z_0} \qquad (7.104)$$

$$\rho_G = \frac{Z_G - Z_0}{Z_G + Z_0} \qquad (7.105)$$

Substituting Eqs. 7.103–7.105 in Eq. 7.102, we obtain an amplitude at load that asymptotically approaches V_L as

$$V_P^\infty = \frac{Z_L V}{Z_L + Z_G} = V_L \qquad (7.106)$$

Equation 7.106 indicates that when we wait long enough, any "transmission line" effects should go away, and we converge to what we would have if the line was just some wire connecting the source to the load, as shown in Fig. 7.12.

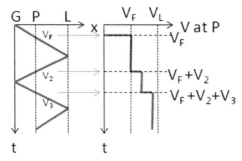

Figure 7.12 Transmission line effect.

On the other hand, if a smooth joint exists between two strings of different impedances, energy will be reflected at the boundary. However, insertion of a particular length of another string between these two *mismatched* strings eliminates energy reflection and matches the impedances, as shown in Fig. 7.13.

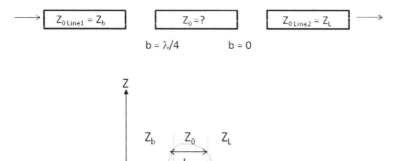

Figure 7.13 Quarter wave transformer.

The general expression of impedance at x is

$$Z(x) = \frac{V(x)}{I(x)} \tag{7.107}$$

Remember that

$$V(x) = V_F e^{-ikx} + V_B e^{ikx} \tag{7.108}$$

$$I(x) = I_F e^{-ikx} + I_B e^{ikx} \tag{7.109}$$

Substituting Eqs. 7.108 and 7.109 in Eq. 7.107, we get

$$Z(x) = \frac{V_F e^{-ikx} + V_B e^{ikx}}{I_F e^{-ikx} + I_B e^{ikx}} \tag{7.110}$$

$$Z(x) = \frac{V_F e^{-ikx} + V_B e^{ikx}}{\left(\dfrac{V_F}{Z_0}\right) e^{-ikx} - \left(\dfrac{V_B}{Z_0}\right) e^{ikx}} \tag{7.111}$$

$$Z(x) = Z_0 \left[\frac{e^{-ikx} + \left(\dfrac{V_B}{V_F}\right) e^{ikx}}{e^{-ikx} - \left(\dfrac{V_B}{V_F}\right) e^{ikx}} \right] \tag{7.112}$$

$$Z(x) = Z_0 \left[\frac{e^{-ikx} + \left(\dfrac{Z_L - Z_0}{Z_L + Z_0}\right) e^{ikx}}{e^{-ikx} - \left(\dfrac{Z_L - Z_0}{Z_L + Z_0}\right) e^{ikx}} \right] \tag{7.113}$$

$$Z(x) = Z_0 \left[\frac{(Z_L + Z_0)e^{-ikx} + (Z_L - Z_0)e^{ikx}}{(Z_L + Z_0)e^{-ikx} - (Z_L - Z_0)e^{ikx}} \right] \tag{7.114}$$

Remembering that

$e^{ikx} = \cos(kx) + i\sin(kx)$, $e^{-ikx} = \cos(kx) - i\sin(kx)$,

$\cos(-x) = \cos(x)$, and $\sin(-x) = -\sin(x)$,

we can then replace the exponential with sine and cosine and substitute $x = -b$ in Eq. 7.114 and get

$$Z_b = Z(-b) = Z_0 \left[\frac{Z_L \cos(kb) + iZ_0 \sin(kb)}{Z_0 \cos(kb) + iZ_L \sin(kb)} \right] = Z_0 \left[\frac{Z_L + iZ_0 \tan(kb)}{Z_0 + iZ_L \tan(kb)} \right] \tag{7.115}$$

In the case of a quarter of a wavelength back from the load $b = \lambda/4$ and remembering that $k = 2\pi/\lambda$, we get $kb = \pi/2$. We then reach

$$Z_b = Z_0 \left[\frac{Z_L + iZ_0 \tan(\pi/2)}{Z_0 + iZ_L \tan(\pi/2)} \right] \tag{7.116}$$

Impedance matching at point $b = \lambda/4$ finally is

$$Z_b = \frac{Z_0^2}{Z_L} \tag{7.117}$$

This expression is important when we try to connect two lines with different impedances, and we do not want to have any reflections. This leads to the concept of *quarter wave transformer*.

7.5 Electromagnetic Waves

Electromagnetic waves are transverse waves that can travel through space where matter is not present, as shown in Fig. 7.14.

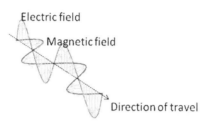

Figure 7.14 Model of electromagnetic waves.

Instead of transferring energy from particle to particle, electromagnetic waves travel by transferring energy between vibrating electric and magnetic fields. The fields regenerate each other. Magnetic and electric fields of electromagnetic waves are perpendicular to each other and to the direction of the waves. All electromagnetic waves travel at 300,000 km/s (velocity of light) in the vacuum of space. The electromagnetic waves travel slower than light when they are in different media, as shown in Table 7.2.

Table 7.2 Velocity of electromagnetic waves in various materials

Materials	Velocity (km/s)
Vacuum	300,000
Air	< 300,000
Water	226,000
Glass	200,000
Diamond	124,000

The denser the medium, the slower the electromagnetic waves travel.

We recall the electromagnetic waves related to the four Maxwell's equations as follows:

Faraday's law (Eq. 1.24): $\quad \vec{\nabla} \times \vec{E} = -\dot{\vec{B}}$

Ampere's law (Eq. 1.25): $\quad \vec{\nabla} \times \vec{H} = \vec{J} + \dot{\vec{D}}$

Gauss's law (electric field) (Eq. 1.26):

$$\vec{\nabla} \cdot \vec{D} = \rho_{\text{TRUE}}$$

Gauss's law (magnetic field) (Eq. 1.27):

$$\vec{\nabla} \cdot \vec{B} = 0$$

where $\dot{\vec{B}} = \dfrac{\partial \vec{B}}{\partial t}$ and $\dot{\vec{D}} = \dfrac{\partial \vec{D}}{\partial t}$

Faraday's law implies that a changing magnetic flux $\dot{\vec{B}}$ has $\vec{E}$ rings around it (Fig. 7.15). Ampere's law implies that a steady current $\vec{J}$ or a changing electric flux $\dot{\vec{D}}$ has magnetic $\vec{H}$ rings around it.

Figure 7.15 Faraday's law (left), Ampere's law (middle), and Gauss's law (right).

Gauss's law implies that $\vec{B}$ flux lines are continuous. The flux lines exiting the north pole of a magnet return to that magnet at the south pole. The $\vec{D}$ flux lines are continuous except when broken by point charges, which means lines of $\vec{D}$ begin and end on point charges.

Recall that $\vec{D}$ and $\vec{E}$, $\vec{B}$ and $\vec{H}$ are directly related by permittivity ε and permeability μ as

$$\vec{D} = \varepsilon \vec{E} \text{ (which is Eq. 2.1)}$$

$$\vec{B} = \mu \vec{H} \text{ (which is Eq. 2.2)}$$

Here $\varepsilon = \varepsilon_0 \, \varepsilon_r$ and ε_0 = permittivity of free space = 8.854×10^{-12} F/m

$\mu = \mu_0 \, \mu_r$ and μ_0 = permeability of free space = $4\pi \times 10^{-7}$ H/m

If we consider a plane wave propagating in a dielectric medium ($\vec{J}$, $\rho = 0$) or lossless medium, Ampere's law from Eq. 1.25 is modified as

$$\vec{\nabla} \times \vec{H} = \frac{\partial \vec{D}}{\partial t} = \varepsilon \frac{\partial \vec{E}}{\partial t} \tag{7.118}$$

By taking curl in Eq. 1.24, we get

$$\vec{\nabla} \times \vec{\nabla} \times \vec{E} = -\vec{\nabla} \times \frac{\partial \vec{B}}{\partial t} = -\mu \vec{\nabla} \times \frac{\partial \vec{H}}{\partial t} \tag{7.119}$$

Substituting Eq. 7.118 in Eq. 7.119, we get

$$\vec{\nabla}(\vec{\nabla}\vec{E}) - \vec{\nabla}^2 \vec{E} = -\mu\varepsilon \frac{\partial^2 \vec{E}}{\partial t^2} \tag{7.120}$$

$$\vec{\nabla}^2 \vec{E} = \mu\varepsilon \frac{\partial^2 \vec{E}}{\partial t^2} \quad \text{(wave equation for electric field)} \tag{7.121}$$

Here a Laplacian operator

$$\Delta = \vec{\nabla} \cdot \vec{\nabla} = \vec{\nabla}^2 \equiv \frac{\partial^2}{\partial x^2} + \frac{\partial^2}{\partial y^2} + \frac{\partial^2}{\partial z^2} \tag{7.122}$$

and

$$v = \sqrt{\frac{1}{\mu\varepsilon}} = 3 \times 10^8 \text{ m/s (light velocity)} \tag{7.123}$$

By taking curl in Eq. 7.118, we get

$$\vec{\nabla} \times \vec{\nabla} \times \vec{H} = \varepsilon \vec{\nabla} \times \frac{\partial \vec{E}}{\partial t} \tag{7.124}$$

Substituting Eq. 1.24, we get

$$\vec{\nabla}(\vec{\nabla}\vec{H}) - \vec{\nabla}^2 \vec{H} = -\varepsilon \frac{\partial^2 \vec{B}}{\partial t^2} = -\mu\varepsilon \frac{\partial^2 \vec{H}}{\partial t^2} \tag{7.125}$$

$$\vec{\nabla}^2 \vec{H} = \mu\varepsilon \frac{\partial^2 \vec{H}}{\partial t^2} \quad \text{(wave equation for magnetic field)} \tag{7.126}$$

The solutions to Eqs. 7.121 and 7.126 are

$$\vec{E}_x = \vec{E}_{xi}\, e^{-i(kz-\omega t)} + \vec{E}_{xr}\, e^{i(kz+\omega t)} \tag{7.127}$$

$$\vec{H}_y = \vec{H}_{yi}\, e^{-i(kz-\omega t)} + \vec{H}_{yr}\, e^{i(kz+\omega t)} \tag{7.128}$$

Since $\vec{E}_{xi}\, e^{-i(kz-\omega t)}$ and $\vec{E}_{xr}\, e^{i(kz+\omega t)}$ in Eq. 7.127 represent waves traveling in opposite direction, they can be treated separately. Therefore, we try a solution for E in Eq. 7.127 of the simplified form as

$$\vec{E}_x = \vec{E}_{xr}\, e^{i(kz+\omega t)} \tag{7.129}$$

and substituting Eq. 7.129 in Eq. 7.121, we get

$$\vec{\nabla}^2 \vec{E} = \mu\varepsilon \frac{\partial^2 \vec{E}}{\partial t^2} \quad \text{(which is Eq. 7.121)}$$

$$(ik)^2 \vec{E}_x = \mu\varepsilon(i\omega)^2 \vec{E}_x \tag{7.130}$$

Then, the phase constant is

$$k = \omega\sqrt{\mu\varepsilon} = \frac{\omega}{v} \tag{7.131}$$

Next, we consider electromagnetic waves propagating in a conductive medium (conductivity σ) or lossy medium ($\vec{J}, \rho \neq 0$). Ampere's law from Eq. 1.25 is modified as

$$\vec{\nabla} \times \vec{H} = \sigma\vec{E} + \frac{\partial \vec{D}}{\partial t} \tag{7.132}$$

By taking curl in Eq. 7.132, we get

$$\vec{\nabla} \times \vec{\nabla} \times \vec{H} = \sigma\vec{\nabla} \times \vec{E} + \vec{\nabla} \times \frac{\partial \vec{D}}{\partial t} \tag{7.133}$$

$$\vec{\nabla}(\vec{\nabla}\vec{H}) - \vec{\nabla}^2 \vec{H} = -\sigma\frac{\partial \vec{B}}{\partial t} + \varepsilon\vec{\nabla} \times \frac{\partial \vec{E}}{\partial t} \tag{7.134}$$

$$-\vec{\nabla}^2 \vec{H} = -\sigma\frac{\partial \vec{B}}{\partial t} - \varepsilon\frac{\partial^2 \vec{B}}{\partial t^2} \tag{7.135}$$

$$\vec{\nabla}^2 \vec{H} = \mu\sigma\frac{\partial \vec{H}}{\partial t} + \mu\varepsilon\frac{\partial^2 \vec{H}}{\partial t^2} \tag{7.136}$$

Defining the magnetic field in the *y*-axis and traveling in the *z*-axis, we get

$$\frac{\partial^2 \vec{H}_y}{\partial z^2} = \mu\sigma\frac{\partial \vec{H}_y}{\partial t} + \mu\varepsilon\frac{\partial^2 \vec{H}_y}{\partial t^2} \tag{7.137}$$

Substituting Eq. 7.128 in Eq. 7.137, we get

$$\frac{\partial^2 \vec{H}_y}{\partial z^2} = (i\omega)\mu\sigma\vec{H}_y + \mu\varepsilon(i\omega)^2 \vec{H}_y \tag{7.138}$$

$$(ik')^2 \vec{H}_y = i\omega\mu(\sigma + i\varepsilon\omega)\vec{H}_y \tag{7.139}$$

and define a real term α corresponding to the attenuation along the line, known as the *attenuation* or *absorption coefficient*, by

$$\alpha + ik = ik' = \sqrt{i\omega\mu(\sigma + i\varepsilon\omega)} \tag{7.140}$$

Replacing k by k' in Eqs. 7.127 and 7.128, we get

$$\vec{E}_x(z,t) = \vec{E}_{xi}\, e^{-i(k'z-\omega t)} + \vec{E}_{xr}\, e^{i(k'z+\omega t)} \tag{7.141}$$

$$\vec{H}_y(z,t) = \vec{H}_{yi}\, e^{-i(k'z-\omega t)} + \vec{H}_{yr}\, e^{i(k'z+\omega t)} \tag{7.142}$$

Substituting $\alpha + ik$, we get

$$\vec{E}_x(z,t) = \vec{E}_{xi}\, e^{-\alpha z-i(kz-\omega t)} + \vec{E}_{xr}\, e^{\alpha z+i(kz+\omega t)} \tag{7.143}$$

$$\vec{H}_y(z,t) = \vec{H}_{yi}\, e^{-\alpha z-i(kz-\omega t)} + \vec{H}_{yr}\, e^{\alpha z+i(kz+\omega t)} \tag{7.144}$$

For the incident electric field, we have (Fig. 7.16)

$$\vec{E}_x(z,t) = \vec{E}_{xi}\, e^{\alpha z-i(kz-\omega t)} \tag{7.145}$$

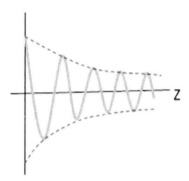

Figure 7.16 Incident electric field.

Defining the magnetic field in the y-axis (electric field in the x-axis) and traveling in the z-axis, Faraday's law is modified as

$$\frac{\partial \vec{E}_x}{\partial z} = -\mu \frac{\partial \vec{H}_y}{\partial t} \tag{7.146}$$

$$-ik\vec{E}_{xi}\, e^{-i(kz-\omega t)} + ik\vec{E}_{xr}\, e^{i(kz+\omega t)} = -i\omega\mu\vec{H}_{yi}\, e^{-i(kz-\omega t)} - i\omega\mu\vec{H}_{yr}\, e^{i(kz+\omega t)} \tag{7.147}$$

Since $e^{-i(kz-\omega t)}$ and $e^{-i(kz+\omega t)}$ represent the waves traveling in opposite directions, they can be treated separately. This leads to two independent expressions in $\vec{E}$ and $\vec{H}$.

$$-ik\vec{E}_{xi}\, e^{-i(kz-\omega t)} = -i\omega\mu\vec{H}_{yi}\, e^{-i(kz-\omega t)} \tag{7.148}$$

$$\frac{\vec{E}_{xi}}{\vec{H}_{yi}} = \frac{\omega\mu}{k} \tag{7.149}$$

$$ik\vec{E}_{xr}\,e^{i(kz+\omega t)} = -i\omega\mu\vec{H}_{yr}\,e^{i(kz+\omega t)} \tag{7.150}$$

$$\frac{\vec{E}_{xr}}{\vec{H}_{yr}} = -\frac{\omega\mu}{k} \tag{7.151}$$

Substituting $k = \omega\sqrt{\mu\varepsilon}$ in Eq. 7.150, we get

$$\frac{\vec{E}_{xi}}{\vec{H}_{yi}} = \sqrt{\frac{\mu}{\varepsilon}} \tag{7.152}$$

If the *intrinsic impedance* η_0 is defined as the ratio of $\vec{E}$ and $\vec{H}$, we get (in a dielectric medium)

$$\eta_0 = \frac{\vec{E}_{xi}}{\vec{H}_{yi}} = -\frac{\vec{E}_{xr}}{\vec{H}_{yr}} = \sqrt{\frac{\mu}{\varepsilon}} \tag{7.153}$$

$$\eta_0 = \sqrt{(4\pi \times 10^{-7}\,\text{H/m})/(8.854 \times 10^{-12}\,\text{F/m})} = 377\,\Omega$$

Since $Z_0 = \sqrt{\dfrac{L}{C}}$ in a lossless transmission line corresponds to

$$Z_0' = \sqrt{\frac{(R + i\omega L)}{(G + i\omega C)}} \tag{7.154}$$

in a lossy transmission line, then $\eta_0 = \sqrt{\dfrac{\mu}{\varepsilon}}$ in free space corresponds to

$$\eta_0' = \sqrt{i\omega\mu/(\sigma + i\varepsilon\omega)} \tag{7.155}$$

in a non-dielectric medium.

The relation between Eqs. 7.154 and 7.155 can be summarized in Table 7.3 (Fig. 7.17).

Table 7.3   Details of coaxial cable and parallel-plate power lines

Details	Coaxial Cable	Parallel-plate power line
Capacitance/length	$C = \dfrac{2\pi\varepsilon}{\ln(b/a)}$	$C = \dfrac{\varepsilon w}{t}$
Inductance/length	$L = \dfrac{\mu}{2\pi}\ln(b/a)$	$L = \dfrac{\mu t}{w}$

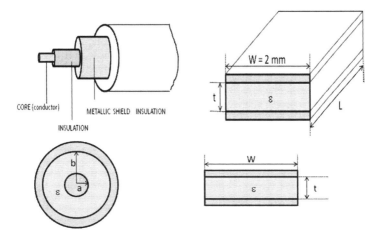

Figure 7.17   Common transmission line structures: coaxial cable (left) and parallel-plate power lines.

Next we will consider reflection of electromagnetic waves in two cases: reflection at normal incident angle and reflection at tilt angle. Reflection at normal incident angle is shown in Fig. 7.18. Suppose that electric and magnetic fields are continuous at the medium interface and

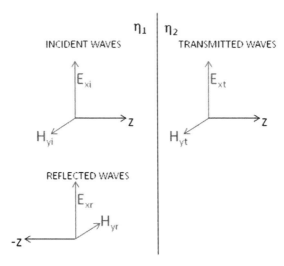

Figure 7.18   Reflection of electromagnetic waves at normal angle.

$$\vec{E}_x(z) = \vec{E}_{xi}\, e^{-ik'z} + \vec{E}_{xr}\, e^{ik'z} \tag{7.156}$$

$$\vec{H}_y(z) = \vec{H}_{yi}\, e^{-ik'z} + \vec{H}_{yr}\, e^{ik'z} \tag{7.157}$$

$$\vec{E}_x(0) = \vec{E}_{xi} + \vec{E}_{xr} = \vec{E}_{xt} \tag{7.158}$$

$$\vec{H}_y(0) = \vec{H}_{yi} + \vec{H}_{yr} = \vec{H}_{yt} \tag{7.159}$$

By definition of η in media 1 and 2, we have

$$\eta_1 = \frac{\vec{E}_{xi}}{\vec{H}_{yi}} = -\frac{\vec{E}_{xr}}{\vec{H}_{yr}} \qquad \eta_2 = \frac{\vec{E}_{xt}}{\vec{H}_{yt}} \tag{7.160}$$

Eliminating $\vec{H}_{yi}$, $\vec{H}_{yr}$, $\vec{H}_{yt}$ using η_1, η_2 gives

$$\frac{\vec{E}_{xi} - \vec{E}_{xr}}{\eta_1} = \frac{\vec{E}_{xt}}{\eta_2} \tag{7.161}$$

Using Eq. 7.158 in Eq. 7.161, we get

$$\frac{\vec{E}_{xi} - \vec{E}_{xr}}{\eta_1} = \frac{\vec{E}_{xi} + \vec{E}_{xr}}{\eta_2} \tag{7.162}$$

Reflection coefficient will be

$$\rho = \frac{\vec{E}_{xr}}{\vec{E}_{xi}} = \frac{\eta_2 - \eta_1}{\eta_2 + \eta_1} \tag{7.163}$$

Reflectance will be

$$R = \rho^2 = \left(\frac{\eta_2 - \eta_1}{\eta_2 + \eta_1}\right)^2 \tag{7.164}$$

If a wave traveling in air strikes a perfect conductor of $\eta_2 = 0$ at normal incidence, then $\rho = \pm 1$ as shown in Fig. 7.23. Thus, good conductors are very good reflectors of electromagnetic waves; for example light waves are well reflected from metal surfaces.

On the other hand, reflection at a tilt incident angle can be explained by Huygens's principle, as shown in Fig. 7.19. Before calculating the reflection coefficient, we need to know the angles at which the refracted and reflected waves will travel by using

$$\frac{CB}{AD} = \frac{v_1}{v_2} = \frac{AB\sin\theta_i}{AB\sin\theta_t} \tag{7.165}$$

$$\frac{\sin\theta_i}{\sin\theta_t} = \sqrt{\frac{\varepsilon_2}{\mu_2} \cdot \frac{\mu_1}{\varepsilon_1}} = n_2 \cdot \frac{1}{n_1} \tag{7.166}$$

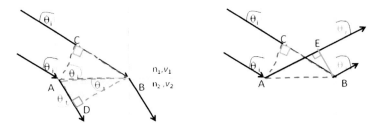

Figure 7.19   Reflection of electromagnetic waves at tilt angle.

Equation 7.166 is *Snell's law* of refraction in which n is the refractive index of the medium.

For a dielectric medium, we assume $\mu_1 = \mu_2 = \mu_0$. Then, we get

$$\frac{\sin\theta_i}{\sin\theta_t} = \sqrt{\frac{\varepsilon_2}{\varepsilon_1}} = \frac{\eta_1}{\eta_2} = \frac{n_2}{n_1} \tag{7.167}$$

By a similar argument, it can be shown that the angle of reflection is equal to the angle of incidence:

$$CB = AE$$

$$AB \sin\theta_i = AB \sin\theta_r$$

$$\theta_i = \theta_r \tag{7.168}$$

In the case of perpendicularly polarized waves (*TE mode*) as shown in Fig. 7.20, a light with perpendicular polarization is called s-polarized (s = senkrecht in German) and we get

$$\vec{E}_i + \vec{E}_r = \vec{E}_t \tag{7.169}$$

$$\vec{B}_i \cos\theta_i + \vec{B}_r \cos\theta_i = \vec{B}_t \cos\theta_t \tag{7.170}$$

$$\vec{H}_i \cos\theta_i + \vec{H}_r \cos\theta_i = \vec{H}_t \cos\theta_t \tag{7.171}$$

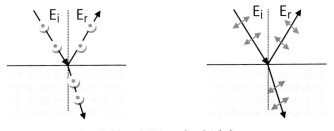

Figure 7.20   TE modes (left) and TM modes (right).

Eliminating $\vec{H}_i$, $\vec{H}_r$, $\vec{H}_t$ using η_1, η_2 gives

$$\frac{\vec{E}_i \cos\theta_i}{\eta_1} - \frac{\vec{E}_r \cos\theta_i}{\eta_1} = \frac{\vec{E}_t \cos\theta_t}{\eta_2} \tag{7.172}$$

$$n_1 \vec{E}_i \cos\theta_i - n_1 \vec{E}_r \cos\theta_i = n_2 \vec{E}_t \cos\theta_t \tag{7.173}$$

$$n_1 \vec{E}_i \cos\theta_i - n_1 \vec{E}_r \cos\theta_i = n_2 (\vec{E}_i + \vec{E}_r) \cos\theta_t \tag{7.174}$$

$$n_1 \cos\theta_i - n_1 \frac{\vec{E}_r}{\vec{E}_i} \cos\theta_i = n_2 \cos\theta_t + n_2 \frac{\vec{E}_r}{\vec{E}_i} \cos\theta_t \tag{7.175}$$

$$n_1 \cos\theta_i - n_2 \cos\theta_t = \frac{\vec{E}_r}{\vec{E}_i} (n_1 \cos\theta_i + n_2 \cos\theta_t) \tag{7.176}$$

$$\rho_\perp = \frac{\vec{E}_r}{\vec{E}_i} = \frac{n_1 \cos\theta_i - n_2 \cos\theta_t}{n_1 \cos\theta_i + n_2 \cos\theta_t} \tag{7.177}$$

Equation 7.177 is *Fresnel's equation* for TE mode.

In the case of parallel polarized waves (*TM mode*) as shown in Fig. 7.20, a light with parallel polarization is called p-polarized (p = parallel) and we get

$$\vec{E}_i \cos\theta_i + \vec{E}_r \cos\theta_i = \vec{E}_t \cos\theta_t \tag{7.178}$$

$$\vec{B}_i + \vec{B}_r = \vec{B}_t \tag{7.179}$$

$$\vec{H}_i + \vec{H}_r = \vec{H}_t \tag{7.180}$$

Eliminating $\vec{H}_i$, $\vec{H}_r$, $\vec{H}_t$ using η_1, η_2 gives

$$\frac{\vec{E}_i}{\eta_1} - \frac{\vec{E}_r}{\eta_1} = \frac{\vec{E}_t}{\eta_2} \tag{7.181}$$

$$n_1 \vec{E}_i - n_1 \vec{E}_r = n_2 \vec{E}_t \tag{7.182}$$

Substituting Eq. 7.182 in Eq. 7.178, we get

$$\vec{E}_i \cos\theta_i + \vec{E}_r \cos\theta_i = \frac{n_1 \vec{E}_i - n_1 \vec{E}_r}{n_2} \cos\theta_t \tag{7.183}$$

$$n_2 \cos\theta_i + n_2 \frac{\vec{E}_r}{\vec{E}_i} \cos\theta_i = n_1 \cos\theta_t - n_1 \frac{\vec{E}_r}{\vec{E}_i} \cos\theta_t \tag{7.184}$$

$$\frac{\vec{E}_r}{\vec{E}_i} (n_2 \cos\theta_i + n_1 \cos\theta_t) = n_1 \cos\theta_t - n_2 \cos\theta_i \tag{7.185}$$

$$\rho_{//} = -\frac{\vec{E}_r}{\vec{E}_i} = \frac{n_2 \cos\theta_i - n_1 \cos\theta_t}{n_2 \cos\theta_i + n_1 \cos\theta_t} \tag{7.186}$$

Equation 7.186 is *Fresnel's equation* for TM mode.

When *unpolarized light* reflects off a transparent dielectric surface, it is partially polarized parallel to the plane of the refractive surface. There is a specific angle called *Brewster's angle* θ_B at which light is 100% polarized. This occurs when the reflected ray and the refracted ray are 90° apart, as shown in Fig. 7.21.

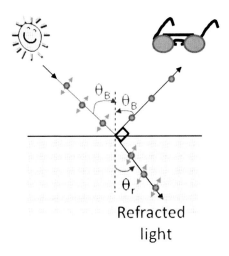

Figure 7.21 Brewster's angle concept.

Since $\theta_r = 90° - \theta_B$ (7.187)

$\sin \theta_r = \sin(90° - \theta_B) = \sin 90° \cos \theta_B - \cos 90° \sin \theta_B = \cos \theta_B$ (7.188)

According to Snell's law

$n_1 \sin \theta_B = n_2 \sin \theta_r$ (7.189)

Substituting Eq. 17.188 in Eq. 7.189, we get

$n_1 \sin \theta_B = n_2 \sin \theta_B$ (7.190)

Thus, we obtain Brewster's angle via

$\tan \theta_B = \dfrac{n_2}{n_1}$ (7.191)

Reflection at Brewster's angle explains why Polaroid films (i.e., sunglasses) cut down reflection, as shown in Fig. 7.22.

Finally, we interpret Fresnel's equations (Eqs. 7.177 and 7.186) in two situations: external and internal reflection. External reflection occurs when a wave moves from a lower-refractive-index medium to a higher one ($n_2/n_1 > 1$), as shown in Fig. 7.23.

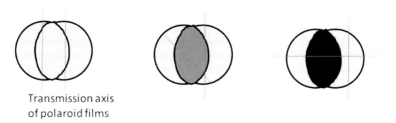

Transmission axis
of polaroid films

Figure 7.22 Reflection by Polaroid films.

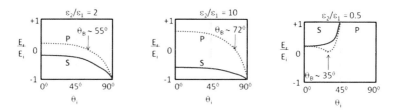

Figure 7.23 External reflection (left and middle) and internal reflection (right).

Brewster's angle tends to increase. The reflection for p-polarized (s-polarized) wave tends to +1 (−1), representing an in-phase (anti-phase) of electromagnetic fields at all angles of incidences. When $n_2/n_1 \to \infty$, total external reflection occurs at all angles of incidences.

Nowadays, optical fiber cables have been introduced instead of Cu axial cables (Fig. 7.24). Compared to traditional metal cables, optical cables have greater and faster data-carrying capacity (bps). In addition, they are more transparent, lighter, and longer.

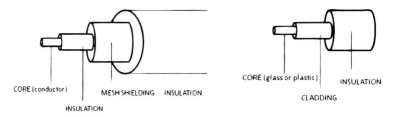

Figure 7.24 Axial cable (left) and optic fiber cable (right).

Total internal reflection occurs when a wave travels in an optical fiber from a higher-refractive-index medium to a lower one ($n_2/n_1 < 1$), as shown in Fig. 7.25. We consider a light with three incident angles passing through the optical fiber (length L with one reflection length L_s).

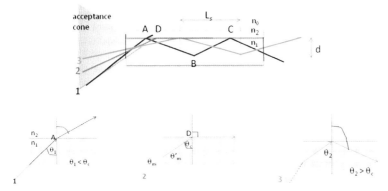

Figure 7.25 Light propagation with different incident angles. θ_c is the critical angle, and n_0, n_1, n_2 are the refractive indices of air, core, and cladding, respectively.

Ray 1 ($\theta_1 < \theta_c$) hits the fiber wall at A at an angle θ_1 where it is partially refracted out of the fiber and partially reflected at B, C, and so on. Ray 2 ($\theta_1 = \theta_c$) entering at an angle θ_m hits the fiber wall at D, where it is refracted parallel to a wall. Ray 3 ($\theta_1 > \theta_c$) entering at $\theta < \theta_m$ meets total internal reflection (no energy loss).

For ray 2,

$$n_0 \sin \theta_m = n_1 \sin \theta'm \quad \text{(Snell's law)} \tag{7.192}$$

$$n_1 \sin \theta_c = n_2 \sin 90° = n_2 \tag{7.193}$$

$$\sin^2 \theta_c = \left(\frac{n_2}{n_1}\right)^2 \tag{7.194}$$

Defining the numerical aperture as

$$NA \equiv n_0 \sin \theta_m \tag{7.195}$$

and applying $\theta'_m = 90° - \theta_c$ and $\sin^2 \theta_c + \cos^2 \theta_c = 1$ in Eq. 7.192, we get

$$NA \equiv n_0 \sin \theta_m = n_1 \sin \theta'_m = n_1 \sin(90° - \theta_c) = n_1 \cos \theta_c \tag{7.196}$$

$$NA = n_1\sqrt{1 - \sin^2\theta_c} \qquad (7.197)$$

Substituting Eq. 7.194 in Eq. 7.197, we have

$$NA = \sqrt{n_1^2 - n_2^2} \qquad (7.198)$$

If d is the fiber diameter (m) and θ_m is the half angle of the largest cone, then $2\theta_m$ is the acceptance cone and θ'_m is the incident half-angled air–cone interface. We can get the number of ray reflection as

$$N = \frac{L}{L_s} = \frac{L}{d\cot\theta'_m} \qquad (7.199)$$

Optical fibers can be categorized into two modes: single mode and multimode (as shown in Fig. 7.26). On one hand, in multimode (i.e. > 50 μm core and > 125 μm clad), many total internal reflections of rays travel into the fiber. This results in large dispersion, spreading the pulses. They are used in computer networks. Multimode can be divided into step index (discontinuous reflection bends rays of light) and grade index (continuous reflection bends rays of light). On the other hand, in single mode (i.e., 5–10 μm core and 125 μm clad), one propagating mode of ray travels into the fiber. This results in less dispersion, more cost of production, and difficulty in coupling to light sources (small NA). They are used in telephone and cable television.

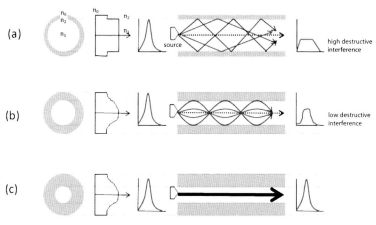

Figure 7.26 Step index (a), grade index (b), and single mode (c).

Performance of optical fibers can be considered from attenuation and dispersion, as shown in Fig. 7.27.

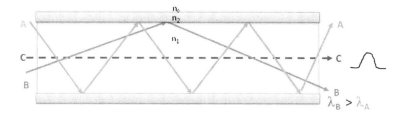

Figure 7.27 Optical fiber performance showing attenuation and dispersion.

On one hand, *attenuation* is the amplitude (E_0) loss of the output pulse in optical fibers mainly due to core absorption. On the other hand, *dispersion* is the broadening output pulse in optical fibers mainly due to modal and chromatic dispersion. That is, modal dispersion is a multiple total internal reflection of rays into fiber. This problem can be reduced by using GRIN fiber or single mode fiber. Chromatic dispersion is not perfect in monochromatic light sources. This problem can be reduced by using light sources of small linewidth.

7.6 Problems

1. Determine the frequency and de Broglie wavelength of a photon having 2 eV energy.
2. What is the momentum of a single photon of red right ($f = 400 \times 10^{12}$ Hz) moving through free space?
3. Determine the de Broglie wavelength corresponding to γ-ray energy of 10^{19} eV.
4. What is the photon energy corresponding to a 60 Hz wave emitted from a power line?
5. A designer is creating a circuit which has a clock rate of 5 MHz and has 200 mm long tracks for which the inductance (L) and capacitance (C) per unit length are 0.5 µH/m and 60 pF/m, respectively. What is the wavelength at 5 MHz?
6. A wave of $V_F = 15$ V with a wavelength $\lambda = 12$ m has a reflected wave of $V_B = 11$ V. If $Z_0 = 75\ \Omega$, what are the voltage and current at the end of the cable?

7. Two lines are to be linked. The first has an impedance $Z_{0 \text{ Line1}}$ = 70 Ω, while the second has an impedance $Z_{0 \text{ Line2}}$ = 150 Ω. What should be the impedance Z_0 of a quarter wavelength section of the line?

8. If we consider a typical b/a = 3 and a typical ε_r = 2.3 for polyethylene materials. Determine the typical value for Z_0.

9. The critical angle for a certain oil is found to be 33°; determine Brewster's angles for this oil.

10. For what refractive index are the critical angle and Brewster's angle equal when the first medium is air?

11. Light is incident upon the air–diamond interface. If the refractive index of diamond is 2.42, determine Brewster's angle and critical angle for this diamond.

12. Determine the reflectance of water (n = 1.33) for both TM and TE modes when the incident angles are 0°, 10°, 45°, and 90°.

13. Unpolarized light is reflected from a plane surface of fused silica glass (n = 1.458). Determine (a) Brewster's angle and critical angle (55.6, 34.4) and (b) reflectance for both TE and TM modes at normal incidence (90°) and at 45°.

14. According to the known free electron concentration, electron path length, and mean velocity as given in the following table, calculate the plasma energy and dielectric constant for each metal and compared with the measured one.

Elements	Sodium	Copper	Silver	Gold
Free electron concentration N (1/cc)	2.65×10^{22}	8.45×10^{22}	5.85×10^{22}	5.90×10^{22}
Plasma energy $\hbar\omega_p$ (eV)	5.71	9.30	9.00	8.50
Electron path length L (nm)	34	42	52	42
Electron mean velocity v_F (cm/s)	1.07×10^8	1.57×10^8	1.39×10^8	1.39×10^8
Dielectric constant at 532 nm	N/A	$-5.5 + i\,5.76$	$-11.78 + i\,0.37$	$-4.71 + i\,2.42$

15. Electron plasma frequency is given by

$$\omega_p^2 = \frac{Ne^2}{m\varepsilon_0}$$

Show that for an electron number density $N \sim 10$ (10^{-5} for atmosphere), electromagnetic waves must have wavelength $\lambda < 3 \times 10^{-3}$ m (in the microwave region) to propagate. These are typical wavelengths for probing thermonuclear plasmas at high temperatures.

$\varepsilon_0 = 8.85 \times 10^{-12}$ F/m; $m = 9.1 \times 10^{-31}$ kg; $e = 1.6 \times 10^{-19}$ C

Chapter 8

Classical and Quantum Theory of Light

8.1 Introduction

In this chapter, we will gain insight into Maxwell's equations under vacuum or even in a medium. We will then study the quantization of light fields in free space.

8.2 Electromagnetic Waves in Vacuum

Recall the following equations:

$$\dot{B} = \frac{\partial \vec{B}}{\partial t} \tag{8.1}$$

$$\dot{D} = \frac{\partial \vec{D}}{\partial t} \tag{8.2}$$

Here, ε_0 ($= 8.854 \times 10^{-12}$ F/m) represents the electric permittivity of vacuum in which

$$\frac{1}{4\pi\varepsilon_0} = 9.0 \times 10^9 \left[\frac{\text{Vm}}{\text{C}} \right] \tag{8.3}$$

Also μ_0 ($= 1.257 \times 10^{-6}$ H/m) represents the magnetic permeability of vacuum in which

$$\frac{1}{\varepsilon_0 \mu_0} = c^2 \tag{8.4}$$

Optical Properties of Solids: An Introductory Textbook
Kitsakorn Locharoenrat
Copyright © 2016 Pan Stanford Publishing Pte. Ltd.
ISBN 978-981-4669-06-1 (Hardcover), 978-981-4669-07-8 (eBook)
www.panstanford.com

With no charges and current present, Maxwell's equations in MKS units are modified as

$$\vec{\nabla} \times \vec{E} = -\dot{\vec{B}} \tag{8.5}$$

$$\vec{\nabla} \times \vec{B} = \varepsilon_0 \mu_0 \dot{\vec{E}} \tag{8.6}$$

$$\vec{\nabla} \cdot \vec{E} = 0 \tag{8.7}$$

$$\vec{\nabla} \cdot \vec{B} = 0 \tag{8.8}$$

Here,

$$\vec{D} = \varepsilon_0 \vec{E} \tag{8.9}$$

$$\vec{B} = \mu_0 \vec{H} \tag{8.10}$$

For a wave equation in isotropic medium under an electric field, we get

$$\vec{\nabla} \times \vec{E} = -\dot{\vec{B}} \tag{8.11}$$

$$\vec{\nabla} \times (\vec{\nabla} \times \vec{E}) = -\frac{\partial}{\partial t} \vec{\nabla} \times \vec{B} \tag{8.12}$$

$$\vec{\nabla} \cdot (\vec{\nabla} \cdot \vec{E}) - (\vec{\nabla} \cdot \vec{\nabla}) \cdot \vec{E} = -\frac{\partial}{\partial t} \left(\varepsilon_0 \mu_0 \frac{\partial \vec{E}}{\partial t} \right) \tag{8.13}$$

$$-(\vec{\nabla} \cdot \vec{\nabla}) \cdot \vec{E} = -\frac{1}{c^2} \frac{\partial^2 \vec{E}}{\partial t^2} \tag{8.14}$$

The wave equation for an electric field is

$$\vec{\nabla}^2 \vec{E} = \frac{1}{c^2} \frac{\partial^2 \vec{E}}{\partial t^2} \tag{8.15}$$

On the other hand, under a magnetic field, we get

$$\vec{\nabla} \times \vec{B} = \varepsilon_0 \mu_0 \dot{\vec{E}} \tag{8.16}$$

$$\vec{\nabla} \times (\vec{\nabla} \times \vec{B}) = \frac{1}{c^2} \frac{\partial}{\partial t} \vec{\nabla} \times \vec{E} \tag{8.17}$$

$$\vec{\nabla} \cdot (\vec{\nabla} \cdot \vec{B}) - (\vec{\nabla} \cdot \vec{\nabla}) \cdot \vec{B} = -\frac{1}{c^2} \frac{\partial}{\partial t} \left(\frac{\partial \vec{B}}{\partial t} \right) \tag{8.18}$$

$$-(\vec{\nabla} \cdot \vec{\nabla}) \cdot \vec{B} = -\frac{1}{c^2} \frac{\partial^2 \vec{B}}{\partial t^2} \tag{8.19}$$

The wave equation for a magnetic field is

$$\vec{\nabla}^2 \vec{B} = \frac{1}{c^2} \frac{\partial^2 \vec{B}}{\partial t^2} \tag{8.20}$$

The plane wave solutions to wave equations in Eqs. 8.15 and 8.20 are

$$\vec{E}(\vec{r},t) = \vec{E}_0 \cos(\vec{k} \cdot \vec{r} - \omega t + \phi) \tag{8.21}$$

$$\vec{B}(\vec{r},t) = \vec{B}_0 \cos(\vec{k} \cdot \vec{r} - \omega t + \phi) \tag{8.22}$$

Here, E_0 and B_0 are the amplitudes of the electric field and the magnetic field, respectively. The angular frequency is $\omega = 2\pi\upsilon$, and ϕ is the initial phase. $\vec{k}$ is the wave vector with magnitude $k = \omega/c$ and direction in the direction of the wave front.

$\vec{E}, \vec{B}$, and $\vec{k}$ form a right-handed system. Electromagnetic waves are transverse waves. They are related to the Poynting vector $\vec{S}$ (the flow of electromagnetic energy transferred into a certain area, as shown in Fig. 8.1) by

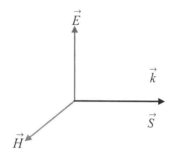

Figure 8.1 Field vectors of light.

$$\vec{S} \equiv \vec{E} \times \vec{H} \tag{8.23}$$

Taking divergence of the Poynting vector $\vec{S}$, we get

$$\vec{\nabla} \cdot \vec{S} = \vec{\nabla} \cdot (\vec{E} \times \vec{H}) \tag{8.24}$$

$$\vec{\nabla} \cdot \vec{S} = \vec{H} \cdot (\vec{\nabla} \times \vec{E}) - \vec{E} \cdot (\vec{\nabla} \times \vec{H}) \tag{8.25}$$

$$\vec{\nabla} \cdot \vec{S} = \left(\vec{H} \cdot (-\dot{\vec{B}}) \right) - \left(\vec{E} \cdot (\dot{\vec{D}}) \right) \tag{8.26}$$

$$\vec{\nabla} \cdot \vec{S} = -\frac{\partial}{\partial t} \left(\frac{1}{2} \vec{E} \cdot \vec{D} - \frac{1}{2} \vec{B} \cdot \vec{H} \right) = -\frac{\partial U}{\partial t} \tag{8.27}$$

The Poynting vector $\vec{S}$ is then related to the electromagnetic energy through the equation of continuity:

$$\vec{\nabla} \cdot \vec{S} + \frac{\partial U}{\partial t} = 0 \tag{8.28}$$

Here, the electromagnetic energy per unit volume U can be defined as

$$U = \frac{1}{2}\vec{E} \cdot \vec{D} + \frac{1}{2}\vec{B} \cdot \vec{H} \tag{8.29}$$

Therefore, the cycled-averaged intensity of the Poynting vector $\vec{S}$ of the plane wave electromagnetic wave will be (for more details, see Appendix A)

$$\bar{I} \equiv \left|\vec{S}\right| = \frac{1}{2}\varepsilon_0\, c \left|\vec{E}_0\right|^2 \tag{8.30}$$

8.3 Polarization of Light

Polarization is the behavior of electromagnetic waves in which they oscillate in more than one orientation (as seen in Fig. 8.2), such as in linear, circular, and elliptical directions.

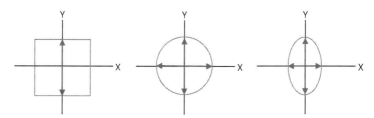

Figure 8.2 Polarization orientations: (left) linear, (center) circular, (right) elliptical.

If the plane wave travels in the z-direction, it can be written as

$$\vec{E}(z,t) = E_x(z,t)\vec{e}_x + E_y(z,t)\vec{e}_y \tag{8.31}$$

The electric field along the x-axis reads

$$E_x(z,t) = E_{x0}\cos(kz - \omega t + \phi_x) \tag{8.32}$$

The electric field along the y-axis reads

$$E_y(z,t) = E_{y0}\cos(kz - \omega t + \phi_y) \tag{8.33}$$

There are so many types of polarizations of light relying on the phase difference in wavelength in the x-and y-directions; the details are summarized in Table 8.1. Note that polarization of natural light occurs in a random phase.

Table 8.1 Polarization of light

Light	Phase
Linearly polarized light	Phase difference $\delta = \phi_y - \phi_x$ $= m\pi \, (m = 0, 1, 2, 3, ...)$
Left circularly polarized light (positive helicity)	$\delta = \left(2m + \dfrac{1}{2} \right)\pi$
Right-circularly polarized light (negative helicity)	$\delta = \left(2m - \dfrac{1}{2} \right)\pi$
Elliptically polarized light	Other phase

An optical filter that allows light with a specific orientation to pass through it is called a *polarizer*. There are two types of polarizers: linear polarizer and circular polarizer. A linear polarizer generates linearly polarized light, whereas a circular polarizer generates circularly polarized light.

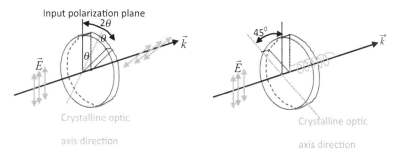

Figure 8.3 Left: A half wave plate reduces about a half cycle the phase difference between the two components of polarized light traversing it. Right: Quarter wave plate is a birefringent material such that the light associated with the larger refractive index is retarded in phase by 90 degree with respect of that associated with the smaller refractive index.

In addition to a polarizer, a *wave plate* is named according to the phase difference in wavelength in x-and y-directions, especially when n_x is not equivalent to n_y. The common types of wave plates are half wave plates (or o $\lambda/2$ wave plates) and quarter wave plates (or $\lambda/4$ wave plates), as shown in Fig. 8.3. A half wave plate rotates the polarization, whereas a quarter wave plate generates the circularly polarized light out of the linearly polarized light.

8.4 Spectrum of Light

The spectrum of light is the power distribution of light fields as compared to energy. If a real electric field is written as

$$\vec{E}(\vec{r},t) = \vec{e}E_0 \cos(\vec{k}\cdot\vec{r} - \omega t + \phi) \tag{8.34}$$

we can write the real part of a complex electric field as

$$\tilde{\vec{E}}(\vec{r},t) = \vec{e}\tilde{E}_0 e^{i(\vec{k}\cdot\vec{r} - \omega t)} \tag{8.35}$$

The spectrum of light is then the square of the Fourier-transformed amplitude of the electric fields (Fig. 8.4), such as

$$\tilde{\vec{E}}(t) = \vec{e}\tilde{E}_0 e^{-0.5\gamma t - i\omega_0 t} \tag{8.36}$$

$$I(\omega) \propto \frac{\left|\tilde{E}_0\right|^2}{(\omega - \omega_0)^2 + (0.5\gamma)^2} \tag{8.37}$$

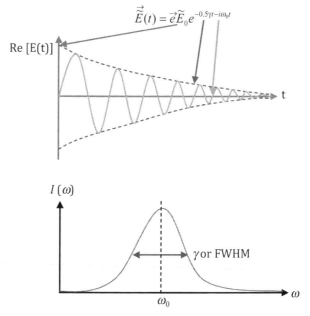

Figure 8.4 Fourier transform of the real part of a complex electric field to get the spectrum of light. γ is the damping constant. FWHM is full width at half maximum.

8.5   Electromagnetic Waves in Medium

With charges and current present, Maxwell's equations in MKS units are written as follows:

$$\vec{\nabla} \times \vec{E} = -\dot{\vec{B}} \tag{8.38}$$

$$\vec{\nabla} \times \vec{H} = \vec{J} + \dot{\vec{D}} \tag{8.39}$$

$$\vec{\nabla} \cdot \vec{D} = \rho_t \tag{8.40}$$

$$\vec{\nabla} \cdot \vec{B} = 0 \tag{8.41}$$

Here, $\vec{D} = \varepsilon \vec{E}$ $\tag{8.42}$

$$\vec{D} = \varepsilon_0 \vec{E} + \vec{P} \tag{8.43}$$

$$\vec{H} = \frac{\vec{B}}{\mu_0} - \vec{M} = \frac{\vec{B}}{\mu} \tag{8.44}$$

$$\vec{J} = \sigma \vec{E} \tag{8.45}$$

where ρ_t, σ, μ, and σ are true charge density, dielectric function, magnetic permeability, and electric conductivity, respectively. $\vec{J}$ is the current density.

For a wave equation in an anisotropic medium, we have the wave equation for the electric field as

$$\vec{\nabla}^2 \vec{E} = \varepsilon\mu \frac{\partial^2 \vec{E}}{\partial t^2} \tag{8.46}$$

and the wave equation for the magnetic field as

$$\vec{\nabla}^2 \vec{B} = \varepsilon\mu \frac{\partial^2 \vec{B}}{\partial t^2} \tag{8.47}$$

Here, refractive index is $n = \sqrt{\dfrac{\varepsilon\mu}{\varepsilon_0\mu_0}}$, and the velocity of light in the medium is $c' = \dfrac{c}{n}$. $\vec{k}$ is the wave vector with magnitude $k = \dfrac{\omega n}{c}$.

An electromagnetic wave is a transverse wave, and it is related to the Poynting vector $\vec{S}$ (a flow of electromagnetic energy) by

$$\vec{S} \equiv \vec{E} \times \vec{H} \tag{8.48}$$

Taking divergence of the Poynting vector $\vec{S}$, we get

$$\vec{\nabla} \cdot \vec{S} = \vec{\nabla} \cdot (\vec{E} \times \vec{H}) \tag{8.49}$$

$$\vec{\nabla} \cdot \vec{S} = \vec{H} \cdot (\vec{\nabla} \times \vec{E}) - \vec{E} \cdot (\vec{\nabla} \times \vec{H}) \tag{8.50}$$

$$\vec{\nabla} \cdot \vec{S} = \left(\vec{H} \cdot (-\dot{\vec{B}}) \right) - \left(\vec{E} \cdot (\vec{J} + \dot{\vec{D}}) \right) \tag{8.51}$$

$$\vec{\nabla} \cdot \vec{S} = -\frac{\partial}{\partial t} \left(\frac{1}{2} \vec{E} \cdot \vec{D} - \frac{1}{2} \vec{B} \cdot \vec{H} \right) - (\vec{J} \cdot \vec{E}) \tag{8.52}$$

If the electromagnetic energy within a unit volume is

$$U = \frac{1}{2} \vec{E} \cdot \vec{D} + \frac{1}{2} \vec{B} \cdot \vec{H} \tag{8.53}$$

then Eq. 8.52 becomes

$$\vec{\nabla} \cdot \vec{S} = -\dot{U} - (\vec{J} \cdot \vec{E}) \tag{8.54}$$

Finally, we obtain the equation of continuity as (the sun is used for analogy in Fig. 8.5)

$$\vec{\nabla} \cdot \vec{S} + \dot{U} + \vec{J} \cdot \vec{E} = 0 \tag{8.55}$$

Here, Joule's heat is $\quad \vec{J} \cdot \vec{E} = \sigma |\vec{E}|^2 \tag{8.56}$

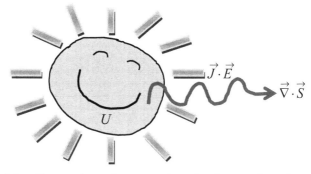

Figure 8.5 The sun is used as an analogy for the equation of continuity. When light propagates, Joule's heat is formed and the electromagnetic energy tends to decrease.

How should the cycled-averaged intensity of the Poynting vector of an electromagnetic wave in a medium be?

8.6   Electromagnetic Potential

The $\vec{\nabla} \cdot B = 0$ in Maxwell's equation will automatically satisfy if we define the scalar potential $\phi(\vec{r}, t)$ and the vector potential $\vec{A}(\vec{r}, t)$ as

$$\vec{B} = \vec{\nabla} \times \vec{A} \tag{8.57}$$

From Faraday's law (Eq. 8.38), we have

$$\vec{\nabla} \times \vec{E} = -\frac{\partial B}{\partial t}$$

We substitute Eq. 8.57 in Eq. 8.38 as

$$\vec{\nabla} \times \vec{E} = -\frac{\partial}{\partial t}(\vec{\nabla} \times \vec{A}) \tag{8.58}$$

$$\vec{\nabla} \times \left(\vec{E} + \frac{\partial \vec{A}}{\partial t} \right) = 0 \tag{8.59}$$

This is satisfied identically by a scalar potential if

$$\vec{E} + \frac{\partial \vec{A}}{\partial t} = -\vec{\nabla}\phi \tag{8.60}$$

or $\quad \vec{E} = -\vec{\nabla}\phi - \dfrac{\partial \vec{A}}{\partial t}$ \hfill (8.61)

There is a certain extent of arbitrariness in the definition of $\phi(\vec{r}, t)$ and $\vec{A}(\vec{r}, t)$ as follows:

$$\vec{A} = \vec{A}_0 + \vec{\nabla}u \tag{8.62}$$

$$\phi = \phi_0 - \frac{\partial u}{\partial t} \tag{8.63}$$

where u is some scalar function. This transformation of potentials is called *gauge transformation*. This invariance of fields is called *gauge invariance*.

From Eq. 8.61, we have $\quad \vec{E} = -\vec{\nabla}\phi - \dfrac{\partial \vec{A}}{\partial t}$

Substituting Eqs. 8.62 and 8.63 in Eq. 8.61, we get

$$\vec{E} = -\vec{\nabla}\left(\phi_0 - \frac{\partial u}{\partial t} \right) - \frac{\partial}{\partial t}(\vec{A}_0 + \vec{\nabla}u) \tag{8.64}$$

$$-\vec{\nabla}\phi - \frac{\partial \vec{A}}{\partial t} = -\vec{\nabla}\left(\phi_0 - \frac{\partial u}{\partial t} \right) - \frac{\partial}{\partial t}(\vec{A}_0 + \vec{\nabla}u) \tag{8.65}$$

$$\vec{E} = -\vec{\nabla}\phi - \frac{\partial \vec{A}}{\partial t} = -\vec{\nabla}\phi_0 + \frac{\partial}{\partial t}\vec{\nabla}u - \frac{\partial \vec{A}_0}{\partial t} - \frac{\partial}{\partial t}\vec{\nabla}u \tag{8.66}$$

These show that the electric fields are left unchanged by the transformation of the potentials.

From Eq. 8.39, we have $\quad \vec{\nabla} \times \vec{B} = \mu \vec{J} + \mu \varepsilon \dfrac{\partial \vec{E}}{\partial t}$

Substituting the gauge transformations from Eqs. 8.57 and 8.61 in Eq. 8.39, we get

$$\vec{\nabla} \times (\vec{\nabla} \times \vec{A}) = \mu \vec{J} + \mu \varepsilon \frac{\partial}{\partial t} \left(-\vec{\nabla}\phi - \frac{\partial \vec{A}}{\partial t} \right) \tag{8.67}$$

$$\vec{\nabla}(\vec{\nabla} \cdot \vec{A}) - \vec{\nabla}^2 \vec{A} = \mu \vec{J} - \frac{1}{c^2} \vec{\nabla}\frac{\partial \phi}{\partial t} - \frac{1}{c^2}\frac{\partial^2 \vec{A}}{\partial t^2} \tag{8.68}$$

$$\vec{\nabla}^2 \vec{A} - \frac{1}{c^2}\frac{\partial^2 \vec{A}}{\partial t^2} - \vec{\nabla}\left(\vec{\nabla} \cdot \vec{A} + \frac{1}{c^2}\frac{\partial \phi}{\partial t} \right) = -\mu \vec{J} \tag{8.69}$$

If we define the *Lorenz gauge* condition frequently used for a system with electric charge and current as

$$\vec{\nabla} \cdot \vec{A} + \frac{1}{c^2}\frac{\partial \phi}{\partial t} = 0 \tag{8.70}$$

and apply the Lorenz gauge condition to Eq. 8.69, we get

$$\vec{\nabla}^2 \vec{A} - \frac{1}{c^2}\frac{\partial^2 \vec{A}}{\partial t^2} = -\mu \vec{J} \tag{8.71}$$

On the other hand, from Eq. 8.40, we have

$$\vec{\nabla} \cdot \vec{E} = \frac{\rho_t}{\varepsilon}$$

Substituting Eq. 8.61 in Eq. 8.40, we get

$$\vec{\nabla} \cdot \left(-\vec{\nabla}\phi - \frac{\partial \vec{A}}{\partial t} \right) = \frac{\rho_t}{\varepsilon} \tag{8.72}$$

$$\Delta\phi + \frac{\partial}{\partial t}(\vec{\nabla} \cdot \vec{A}) = -\frac{\rho_t}{\varepsilon} \tag{8.73}$$

And applying the Lorenz gauge condition to Eq. 8.73, we get

$$\Delta\phi - \frac{1}{c^2}\frac{\partial^2 \phi}{\partial t^2} = -\frac{\rho_t}{\varepsilon} \tag{8.74}$$

Use d'Alembert operator (represented by a box: □) as

$$\square \equiv \nabla^2 - \frac{1}{c^2}\frac{\partial^2}{\partial t^2} \equiv \Delta - \frac{1}{c^2}\frac{\partial^2}{\partial t^2} \tag{8.75}$$

Maxwell's equations from Eqs. 8.71 and 8.74 will be

$$\Box \vec{A} = -u\,\vec{J} \tag{8.76}$$

$$\Box \phi = -\frac{\rho_t}{\varepsilon} \tag{8.77}$$

Solution to these Maxwell's equations will be

$$\phi(\vec{r},t) = \frac{1}{4\pi\varepsilon} \int \frac{\rho_t(\vec{r}',t')}{R}\,d\vec{r}' \tag{8.78}$$

$$\vec{A}(\vec{r},t) = \frac{1}{4\pi\varepsilon} \int \frac{\vec{J}(\vec{r}',t')}{R}\,d\vec{r}' \tag{8.79}$$

Here, $t' = t\dfrac{R}{c'}$ and $R \equiv |\vec{r} - \vec{r}'|$

When there is charge and current, together with dielectric, we can apply the Lorenz gauge condition to Eq. 8.69 and get

$$[\vec{\nabla}^2\vec{A} - \frac{1}{c^2}\frac{\partial^2\vec{A}}{\partial t^2} - \vec{\nabla}\left(\vec{\nabla}\cdot\vec{A} + \frac{1}{c^2}\frac{\partial\phi}{\partial t}\right) = -\mu\vec{J} \quad \text{(which is Eq. 8.69)]}$$

$$\vec{\nabla}^2\vec{A} - \frac{1}{c^2}\frac{\partial^2\vec{A}}{\partial t^2} - \vec{\nabla}\,\vec{\nabla}\cdot\vec{A} - \frac{1}{c^2}\vec{\nabla}\frac{\partial\phi}{\partial t} = -\mu\vec{J} \tag{8.80}$$

And we define the *Coulomb gauge* condition normally used in field quantization as

$$\vec{\nabla}\cdot\vec{A} = 0 \tag{8.81}$$

We apply the Coulomb gauge condition to Eq. 8.80 and get

$$\Box\,\vec{A} - \frac{1}{c^2}\vec{\nabla}\frac{\partial\phi}{\partial t} = -\mu\,\vec{J} \tag{8.82}$$

$$\Box\,\vec{A} - \frac{1}{c^2}\vec{\nabla}\frac{\partial\phi}{\partial t} = -\mu\left(-\mu^{-1}*\vec{A} + \varepsilon\vec{\nabla}\frac{\partial\phi}{\partial t}\right) \tag{8.83}$$

If any vector field $\vec{J}$ is expressed as the sum of the longitudinal (irrotational) term and the transverse (solenoidal) term or

$$\vec{J} = \vec{J}_{\text{LONGITUDINAL}} + \vec{J}_{\text{TRANSVERSE}} \tag{8.84}$$

Then, we will get the components of any vector field as

$$\varepsilon\vec{\nabla}\frac{\partial\phi}{\partial t} = \vec{J}_{\text{LONGITUDINAL}} \tag{8.85}$$

$$\Box \ \vec{A} = -\mu \, \vec{J}_{\text{TRANSVERSE}} \tag{8.86}$$

On the other hand, applying the Coulomb gauge condition to Eq. 8.73, we get

$$[\, \Delta\phi + \frac{\partial}{\partial t}(\vec{\nabla} \cdot \vec{A}) = -\frac{\rho_t}{\varepsilon} \qquad \text{(which is Eq. 8.73)]}$$

$$\nabla^2 \phi = -\frac{\rho_t}{\varepsilon} \tag{8.87}$$

8.7 Electromagnetic Field Quantization

The quantization of electromagnetic fields is used to define elementary excitations, as shown in Fig. 8.6.

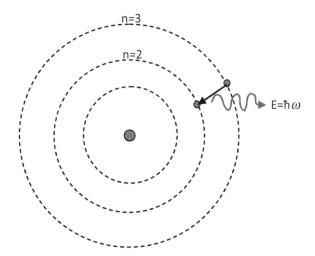

Figure 8.6 Quantized field in free space of a matter.

The vector potential previously explained in Section 8.6 as

$$\vec{A}(\vec{r}, t) = \sum_k \vec{A}_k(t) e^{i\vec{k}\cdot\vec{r}} \tag{8.88}$$

will be quantized in the Coulomb gauge. We impose the periodic boundary conditions in a cube of volume L^3 as follows:

a. Wave vector $k_j = \dfrac{2\pi n_j}{L}$ in which n_j is an integer and $j = (x, y, z)$.

b. $\vec{e}_\lambda \times \vec{h}_\lambda = \dfrac{\vec{k}_\lambda}{|\vec{k}_\lambda|}$ in which $\vec{e}_\lambda$ is polarization.

c. Vector potential $\vec{A}_\lambda(t) = A_\lambda \vec{e}_\lambda \exp(-i\omega_\lambda t)$ in which $\omega_\lambda = |\vec{k}_\lambda| c$.

d. Harmonic oscillator operators $\hat{a}_\lambda = (2\hbar\omega_\lambda)^{-0.5}(\omega_\lambda \hat{Q}_\lambda + i\hat{P}_\lambda)$ and $\hat{a}_\lambda^+ = (2\hbar\omega_\lambda)^{-0.5}(\omega_\lambda \hat{Q}_\lambda - i\hat{P}_\lambda)$.

e. $(\hat{P}_\lambda, \hat{Q}_\lambda)$ are a set of phonon coordinate operators and defined as $P_\lambda = i\omega_\lambda \sqrt{\varepsilon_0 V}(A_\lambda^* - A_\lambda)$ and $Q_\lambda = \sqrt{\varepsilon_0 V}(A_\lambda^* + A_\lambda)$.

Thus, the vector potential and electromagnetic field will be

$$\vec{A}_\lambda = \sqrt{\frac{\hbar}{2\varepsilon_0 V \omega_\lambda}}\, \vec{e}_\lambda \left[\hat{a}_\lambda e^{i(\vec{k}_\lambda \cdot \vec{r} - \omega_\lambda t)} + \hat{a}_\lambda^+ e^{-i(\vec{k}_\lambda \cdot \vec{r} - \omega_\lambda t)}\right] \tag{8.89}$$

$$\vec{E}_\lambda = i\sqrt{\frac{\hbar\omega_\lambda}{2\varepsilon_0 V}}\, \vec{e}_\lambda \left[\hat{a}_\lambda e^{i(\vec{k}_\lambda \cdot \vec{r} - \omega_\lambda t)} - \hat{a}_\lambda^+ e^{-i(\vec{k}_\lambda \cdot \vec{r} - \omega_\lambda t)}\right] \tag{8.90}$$

$$\vec{H}_\lambda = i\sqrt{\frac{\hbar\omega_\lambda}{2\mu_0 V}}\, \vec{h}_\lambda \left[\hat{a}_\lambda e^{i(\vec{k}_\lambda \cdot \vec{r} - \omega_\lambda t)} - \hat{a}_\lambda^+ e^{-i(\vec{k}_\lambda \cdot \vec{r} - \omega_\lambda t)}\right] \tag{8.91}$$

From Eq. 8.91, the phonon Hamiltonian can generally transform in terms of the harmonic oscillators as

$$\hat{H}_\lambda = \hbar\omega_\lambda \left(\hat{a}_\lambda^+ \hat{a}_\lambda + 0.5\right) \tag{8.92}$$

Here, harmonic oscillator operators $\hat{a}_\lambda^+$ and $\hat{a}_\lambda$ are creation operator and annihilation operator, respectively. They also correspond to the classical field amplitudes. The $\hat{a}_\lambda^+$ making one photon is defined as

$$\hat{a}^+ |n\rangle = \sqrt{n+1}\,|n+1\rangle \tag{8.93}$$

The $\hat{a}$ deleting one photon is defined as

$$\hat{a}|n\rangle = \sqrt{n}\,|n-1\rangle \tag{8.94}$$

The $\hat{a}$ and $\hat{a}^+$ both relate to the number operator $\hat{n}$ as

$$\hat{a}^+ \hat{a} \equiv \hat{n} \tag{8.95}$$

8.8 Problems

1. Prove that light is a wave–particle duality according to de Broglie's equation.
2. Prove that the speed of light is 3×10^8 m/s.

3. Prove that the cycled-averaged intensity of the Poynting vector $\vec{S}$ of the plane wave electromagnetic wave in vacuum is

$$I \equiv |\vec{S}| = \frac{1}{2}\varepsilon_0 c |\vec{E}_0|^2$$

4. Show that the complex electric field of circularly polarized plane wave moving along the z-direction is

$$\vec{E}(\vec{r},t) = E_0(\vec{e}_x \pm \vec{e}_y)e^{i(kz-\omega t)}$$

Which sign of $\pm$ corresponds to the right-circularly polarized light? Also show that a linearly polarized light wave is the superposition of two circularly polarized light waves.

5. Describe how a half wave plate operates.

6. Explain the electromagnetic potential and gauge transformation, including Lorentz and Coulomb gauges.

7. Show that the relations

$$\langle n|\hat{E}|n\rangle = 0$$

$$\langle n|\hat{E}^2|n\rangle = \frac{\hbar\omega}{\varepsilon_0 V}(n+0.5)$$

hold for light mode with one degree of freedom. This relation indicates that the strengths of electric and magnetic fields are not zero even if there is no photon, but they fluctuate around the value zero. This fluctuation is called the zero-point vibration.

8. Show that the number of light modes with wave vectors between k and $k + dk$ in a cube of volume L^3 is

$$dN(k) = 2\left(\frac{L}{2\pi}\right)^3 dk_x\, dk_y\, dk_z \text{ where } dk = (dk_x, dk_y, dk_z).$$

9. Show that

$$dN(k) = 2\left(\frac{L}{2\pi}\right)^3 k^2 dk\, d\Omega \text{ where } d\Omega \text{ is the solid angle. Also}$$

show that this relation can be rewritten as $D(k)dk = \left(\frac{k}{\pi}\right)^2 dk$.

10. Find the form of the energy density of radiation with wavelength between λ and $\lambda + d\lambda$ in a vacuum box at temperature T.

Chapter 9

Classical Theory of Light–Matter Interaction I

9.1 Introduction

In this chapter, we will learn the classical treatment of light fields. Many useful theories of optical properties of matter, such as the Kramers–Kronig relation, can be proved in the classical framework.

9.2 Optical Constants

In solid-state materials, electrons are localized near positive ions and free electrons, contributing to electrical conductivity. The interaction of light with materials is reflected by a dielectric function. Scattering and absorption of light occur due to the dielectric response from the material, as shown in Fig. 9.1.

Hence, we describe the effect of free electrons in terms of the dielectric function. Suppose that the effect of the magnetic field of light becomes small and that we have *no true charge*. From one of Maxwell's equations, we have

$$\vec{\nabla} \times \vec{H} = \frac{\partial \vec{D}}{\partial t} + \vec{J} \tag{9.1}$$

$$\vec{\nabla} \times \vec{H} = \frac{\partial}{\partial t}\left(\vec{D} + \int \vec{J}dt\right) \tag{9.2}$$

Optical Properties of Solids: An Introductory Textbook
Kitsakorn Locharoenrat
Copyright © 2016 Pan Stanford Publishing Pte. Ltd.
ISBN 978-981-4669-06-1 (Hardcover), 978-981-4669-07-8 (eBook)
www.panstanford.com

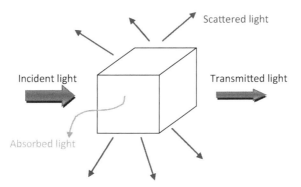

Figure 9.1 Interaction of light with matter.

Using Ohm's law, we get

$$\vec{\nabla} \times \vec{H} = \frac{\partial}{\partial t}\left(\varepsilon \vec{E} - \frac{1}{i\omega}\sigma \vec{E}\right) \tag{9.3}$$

If we define a *complex dielectric function* $\tilde{\varepsilon} = \varepsilon - \frac{1}{i\omega}\sigma$ in which $\vec{D}, \vec{E}, \vec{J} \propto \exp(-i\omega t)$,

Eq. 9.3 becomes

$$\vec{\nabla} \times \vec{H} \equiv \frac{\partial}{\partial t}\tilde{\varepsilon}\,\vec{E} \tag{9.4}$$

This complex dielectric function results in the absorption of light by the material. In addition, when ω reaches zero, we adopt only the real part of $\tilde{\varepsilon}$.

When light travels in a material, the wave equation of the electric field in a *non-magnetic medium* with a complex dielectric function can be written as

$$\vec{\nabla}^2 \vec{E} = \tilde{\varepsilon}\mu_0 \frac{\partial^2 \vec{E}}{\partial t^2} \tag{9.5}$$

and a plane wave solution to the wave equation in Eq. 9.5 is

$$\vec{E}(\vec{r}, t) = \vec{E}_0 e^{i(\vec{k}\cdot\vec{r} - \omega t)} \tag{9.6}$$

A complex wave vector $\vec{k}$ includes the perpendicular to the constant-phase planes $\vec{k}'$ and the constant-amplitude planes $\vec{k}''$ by

$$\vec{k} = \vec{k}' + i\vec{k}'' \tag{9.7}$$

Substituting Eq. 9.6 in Eq. 9.5, we obtain

$$(i\vec{k})^2 \vec{E} = \tilde{\varepsilon}\mu_0 (i\omega)^2 \vec{E} \tag{9.8}$$

$$\vec{\tilde{k}} \cdot \vec{\tilde{k}} = \tilde{\varepsilon}\mu_0\omega^2 \qquad (9.9)$$

Due to

$$\varepsilon_0\mu_0c^2 = 1 \qquad (9.10)$$

substituting Eq. 9.10 in Eq. 9.9, we get

$$\vec{\tilde{k}}^2 = \frac{\tilde{\varepsilon}\omega^2}{\varepsilon_0c^2} \qquad (9.11)$$

If we define the *complex refractive index* as

$$\tilde{n} = \sqrt{\frac{\tilde{\varepsilon}}{\varepsilon_0}} \qquad (9.12)$$

after substituting Eq. 9.12 in Eq. 9.11, we get

$$\vec{\tilde{k}}^2 = \frac{\tilde{n}^2\omega^2}{c^2} \qquad (9.13)$$

From Eqs. 9.12 and 9.13, we can define the *dispersion relation* as

$$\tilde{n} = n + i\kappa = \frac{\vec{\tilde{k}}c}{\omega} = \sqrt{\frac{\tilde{\varepsilon}}{\varepsilon_0}} \qquad (9.14)$$

This dispersion relation is composed of two components: one is the *refractive index* given by

$$n = \frac{\left|\vec{k}'\right|c}{\omega} \qquad (9.15)$$

This refractive index can be found from the refraction angle of light in a prism made of the material in question. Another component of the dispersion relation is the *extinction coefficient* given by

$$\kappa = \frac{\left|\vec{k}''\right|c}{\omega} \qquad (9.16)$$

If the complex dielectric function consists of

$$\tilde{\varepsilon} = \varepsilon' + i\varepsilon'' \qquad (9.17)$$

and we substitute Eq. 9.17 in Eq. 9.14, the following relations also hold between the *optical constants*:

$$\frac{\varepsilon'}{\varepsilon_0} = n^2 - \kappa^2 \qquad (9.18)$$

$$\frac{\varepsilon''}{\varepsilon_0} = 2n\kappa \tag{9.19}$$

$$n^2 = \frac{1}{2\varepsilon_0}\left(\sqrt{\varepsilon'^2 + \varepsilon''^2} + \varepsilon'\right) \tag{9.20}$$

$$\kappa^2 = \frac{1}{2\varepsilon_0}\left(\sqrt{\varepsilon'^2 + \varepsilon''^2} - \varepsilon'\right) \tag{9.21}$$

From Eqs. 9.7 and 9.14, we get

$$n + i\kappa = \frac{\vec{k}c}{\omega} = \left(\vec{k}' + i\vec{k}''\right)\frac{c}{\omega} \tag{9.22}$$

$$\vec{k}' + i\vec{k}'' = \frac{\omega}{c}(n + i\kappa) \tag{9.23}$$

Substituting Eq. 9.23 in Eq. 9.6, we obtain the electric field in the z-direction as

$$\vec{E}(\vec{r},t) = \vec{E}_0 e^{i(\vec{k}'\cdot\vec{r} - \omega t)} e^{-\vec{k}''\cdot\vec{r}} \tag{9.24}$$

$$\vec{E}(z,t) = \vec{E}_0 e^{i\omega\left(\frac{n}{c}z - t\right)} e^{-\frac{\omega\kappa}{c}z} \tag{9.25}$$

The *absorption coefficient* is defined as

$$\alpha = \frac{2\omega\kappa}{c} \tag{9.26}$$

This absorption coefficient can be found from the optical transmittance of a material of known thickness by substituting Eq. 9.26 in Eq. 9.25. Thus, we obtain the electric field in the z-direction as

$$\vec{E}(z,t) = \vec{E}_0 e^{i\omega\left(\frac{n}{c}z - t\right)} e^{-0.5\alpha z} \tag{9.27}$$

The intensity of light is then the square of the electric field (Fig. 9.2) given by

$$I(z) = \left|\vec{E}\right|^2 \tag{9.28}$$

$$I(z) = \left|\vec{E}_0\right|^2 e^{-\alpha z} \tag{9.29}$$

Eq. 9.29 can be rewritten as

$$I_z = I_0 e^{-\alpha z} \tag{9.30}$$

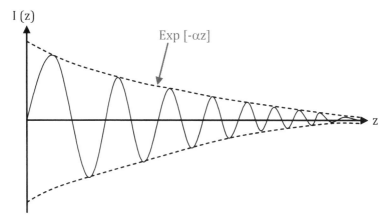

Figure 9.2   Light intensity with respect to the z-direction.

and the optical density will be

$$\ln\left[\frac{I_0}{I_z}\right] = \alpha z \tag{9.31}$$

When a light wave entering a conductor (i.e., metal) is damped while traveling some distance, this is called *skin effect*, as seen in Fig. 9.3.

Lower current density →

Figure 9.3   Skin depth is that distance below the surface at which the current density equals the surface current density multiplied by "1/e."

This means that from the electric field in the z-direction as

$$I(z) = \left|\vec{E}\right|^2 = \left|\vec{E}_0\right|^2 e^{-\alpha z} \tag{9.32}$$

the skin depth at $z = \delta$, at which the amplitude of the wave is damped to $1/e$ from the starting amplitude, is

$$\delta = \frac{1}{\alpha} \tag{9.33}$$

On the other hand, in the case of the magnetic field in a non-magnetic medium, the wave equation with a complex dielectric function is

$$\vec{\nabla}^2\vec{B} = \tilde{\varepsilon}\mu_0 \frac{\partial^2 \vec{B}}{\partial t^2}$$

(9.34)

The plane wave solution to the wave equation in Eq. 9.34 will be as follows:

$$\vec{B}(\vec{r},t) = \vec{B}_0 e^{i(\vec{k}\cdot\vec{r}-\omega t)}$$

(9.35)

$$\vec{B}(\vec{r},t) = \vec{B}_0 e^{i(\vec{k}\cdot\vec{r}-\omega t)} e^{-\vec{k}''\cdot\vec{r}}$$

(9.36)

$$\vec{B}(z,t) = \vec{B}_0 e^{i\omega\left(\frac{n}{c}z-t\right)} e^{-\frac{\omega\kappa}{c}z}$$

(9.37)

9.3 Propagation of Light in a Material

In Fig. 9.4, the boundary conditions of the electromagnetic fields at an interface are divided into two components: (1) tangential part of the electric field E being continuous over a boundary and (2) tangential part of the magnetic field H being continuous over a boundary.

The boundary conditions lead to the relations holding for the tangential components of the wave vector k as follows:

From momentum $\vec{P}$ due to the light traveling on any surface

$$\vec{P} = \hbar\vec{k}$$

(9.38)

together with the conservation of momentum, we get

$$k'_{1//} = k'_{r//} = k'_{2//}$$

(9.39)

when $\quad \theta_r = \theta_1$

(9.40)

From the refractive index (Eq. 9.15), we have

$$n = \frac{\left|\vec{k}'\right|c}{\omega}$$

or $\quad \left|\vec{k}'_1\right| = \frac{n_1\omega}{c}$

(9.41)

And from Snell's law of refraction as

$$n_1 \sin\theta_1 = n_2 \sin\theta_2$$

(9.42)

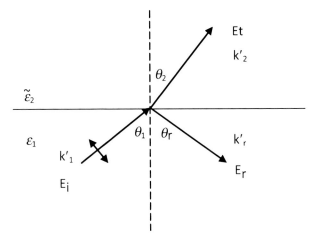

Figure 9.4 Light propagation from medium 1 to medium 2 with dielectric constants ε_1 and $\tilde{\varepsilon}_2$. E_i, E_r, and E_t are incident, reflection, and transmission electric fields, respectively. θ_1, θ_1, and θ_2 are incident, reflection, and transmission angles, respectively. k'_1, k'_r, and k'_2 are incident, reflection, and transmission wave vectors, respectively.

substituting Eq. 9.42 in Eq. 9.41, we obtain

$$\left|\vec{k_1'}\right|\sin\theta_1 = \left|\vec{k_2'}\right|\sin\theta_2 \tag{9.43}$$

In addition, the boundary conditions lead to the relations holding for the amplitudes of the electromagnetic fields, optical reflection and transmission, near a boundary as shown in Fig. 9.5.

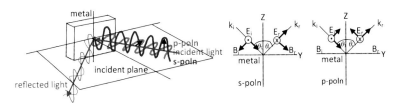

Figure 9.5 Illustration of light on a metal surface. The light polarized perpendicular to the incident plane is called s-polarization, whereas the light polarized parallel to the incident plane is called p-polarization. The incident plane is the plane that contains the incident and reflected lights.

The reflection coefficient $\tilde{r}_\alpha$ and transmission coefficient $\tilde{t}_\alpha$ of α-polarized light at a plane boundary between medium 1 and medium 2 with dielectric constants ε_1 and $\tilde{\varepsilon}_2$, when light travels from medium 1 to medium 2, are

$$\frac{E_{r,s}}{E_{i,s}} = \tilde{r}_s = \tilde{t}_s - 1 = \frac{\cos\theta_1 - \sqrt{\dfrac{\tilde{\varepsilon}_2}{\varepsilon_1} - \sin^2\theta_1}}{\cos\theta_1 + \sqrt{\dfrac{\tilde{\varepsilon}_2}{\varepsilon_1} - \sin^2\theta_1}} \tag{9.44}$$

$$\frac{E_{r,p}}{E_{i,p}} = \tilde{r}_p = \tilde{t}_p - 1 = \frac{\dfrac{\tilde{\varepsilon}_2}{\varepsilon_1}\cos\theta_1 - \sqrt{\dfrac{\tilde{\varepsilon}_2}{\varepsilon_1} - \sin^2\theta_1}}{\dfrac{\tilde{\varepsilon}_2}{\varepsilon_1}\cos\theta_1 + \sqrt{\dfrac{\tilde{\varepsilon}_2}{\varepsilon_1} - \sin^2\theta_1}} \tag{9.45}$$

Figure 9.6 is one of our examples of the reflectivity-polarization analysis using Eqs. 9.44 and 9.45.

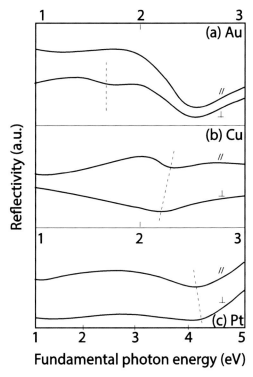

Figure 9.6   Reflectivity of Au (a), Cu (b), and Pt (c) nanowires. ⊥ (//) means the electric field is perpendicular (parallel) to the wire axes.

If $\theta_1 = 0$, Eq. 9.44 becomes

$$\tilde{r}_S = \frac{1 - \sqrt{\dfrac{\tilde{\varepsilon}_2}{\varepsilon_1}}}{1 + \sqrt{\dfrac{\tilde{\varepsilon}_2}{\varepsilon_1}}} \qquad (9.46)$$

$$\tilde{r}_S = \frac{\sqrt{\dfrac{\varepsilon_1}{\varepsilon_0}} - \sqrt{\dfrac{\tilde{\varepsilon}_2}{\varepsilon_1}\dfrac{\varepsilon_1}{\varepsilon_0}}}{\sqrt{\dfrac{\varepsilon_1}{\varepsilon_0}} + \sqrt{\dfrac{\tilde{\varepsilon}_2}{\varepsilon_1}\dfrac{\varepsilon_1}{\varepsilon_0}}} \qquad (9.47)$$

$$\tilde{r}_S = \frac{\sqrt{\dfrac{\varepsilon_1}{\varepsilon_0}} - \sqrt{\dfrac{\tilde{\varepsilon}_2}{\varepsilon_0}}}{\sqrt{\dfrac{\varepsilon_1}{\varepsilon_0}} + \sqrt{\dfrac{\tilde{\varepsilon}_2}{\varepsilon_0}}} \qquad (9.48)$$

From the complex refractive index as

$$\tilde{n} = \sqrt{\frac{\tilde{\varepsilon}}{\varepsilon_0}} \qquad \text{(which is Eq. 9.12)}$$

substituting Eq. 9.12 in Eq. 9.48, we get

$$\tilde{r}_S = \frac{n_1 - n_2}{n_1 + n_2} \qquad (9.49)$$

Equation 9.49 indicates that the phase of the refracted light wave at the boundary between medium 1 and medium 2 is the same as that of the incident wave when refractive index $n_1 > n_2$; however, it is reversed when $n_1 < n_2$ when the light travels from medium 1 to medium 2, as shown in Fig. 9.7.

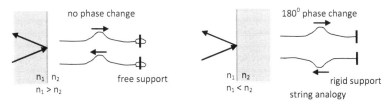

Figure 9.7   Phase change produced by refraction.

For a normal incidence of light as shown in Fig. 9.8, the absolute value of the reflectivity $|\tilde{r}|$ at the boundary between medium 1 and medium 2 of a light wave traveling from medium 1 to medium 2 is the same as that of a light wave traveling from medium 2 to medium 1.

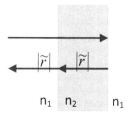

Figure 9.8 Incident light at normal incidence.

When $\tilde{\varepsilon}_2$ is negative (i.e., metal), or when both ε_1 and $\tilde{\varepsilon}_2$ are real and $\varepsilon_1 > \tilde{\varepsilon}_2$, as in the case when light comes out of water, and when the *incident angle* θ_1 is larger than the *critical angle* θ_c (Fig 9.9), *total reflection* occurs. The square root terms of Eqs. 9.44 and 9.45 can be negative, that is

$$\frac{\tilde{\varepsilon}_2}{\varepsilon_1} - \sin^2 \theta_1 \ \langle \ 0 \tag{9.50}$$

$$\frac{\tilde{\varepsilon}_2}{\varepsilon_1} \ \langle \ \sin^2 \theta_1 \tag{9.51}$$

$$\tilde{r} = \frac{\cos\theta_1 - ia}{\cos\theta_1 + ia} \tag{9.52}$$

$$|\tilde{r}| = \frac{\sqrt{\cos^2 \theta_1 - a^2}}{\sqrt{\cos^2 \theta_1 - a^2}} = 1 \tag{9.53}$$

This means that the absolute value of the reflection coefficient $|\tilde{r}|$ is unity.

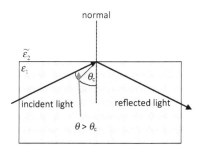

Figure 9.9 The incident angle θ_1 of light is larger than the critical angle θ_c.

In Fig. 9.10, when total reflection happens, a phase shift and a shift of the emission portion for the reflected light are observed at the reflection. This is called the *Goos-Hänchen effect*.

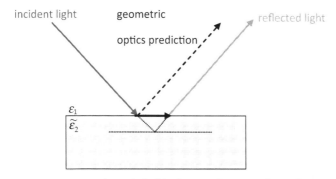

Figure 9.10   Goos–Hänchen shift (black arrow) in metallic reflection. The spatial and angular shifts are exaggerated.

9.4   Kramers–Kronig Relations

We define the electric field coefficient $\vec{E}(\omega)$ as

$$\vec{E}(t) = \int_{-\infty}^{\infty} \vec{E}(\omega)\, e^{-i\omega t}\, d\omega \qquad (9.54)$$

Using Fourier transform, we obtain

$$\vec{E}(\omega) = \frac{1}{2\pi} \int_{-\infty}^{\infty} \vec{E}(t)\, e^{i\omega t}\, dt \qquad (9.55)$$

Also we define the electric susceptibility $\chi(\omega)$ through the relations

$$\vec{D}(\omega) = \varepsilon(\omega)\vec{E}(\omega) \qquad (9.56)$$

$$\vec{D}(\omega) = \varepsilon_0 \vec{E}(\omega) + \vec{P}(\omega) \qquad (9.57)$$

$$\vec{D}(\omega) = \varepsilon_0 \vec{E}(\omega) + \varepsilon_0\, \chi(\omega)\vec{E}(\omega) \qquad (9.58)$$

Thus, $\varepsilon(\omega) = \varepsilon_0 \left[1 + \chi(\omega) \right]$ \qquad (9.59)

General mathematical relations hold between the real and imaginary elements of the holomorphic complex function (Fig. 9.11).

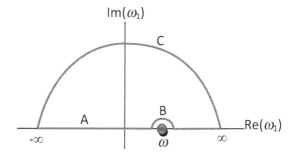

Figure 9.11 Contour integration of holomorphic complex function.

When this complex function is an optical response function, the relations are called *Kramers–Kronig (K–K) relations*. The real and imaginary components of electric susceptibility $\chi(\omega)$ are then related to each other through the relations

$$\chi(\omega) = \chi'(\omega) + i\chi''(\omega) \tag{9.60}$$

$$\oint \frac{\chi(\omega_1)}{\omega_1 - \omega} d\omega_1 = 0 = \int_{(1)} + \int_{(2)} + \int_{(3)} \tag{9.61}$$

With the property of $\chi(\omega)$ having no pole beyond the x-direction, these three segments can be defined as

$$\int_{(1)} = P \int_{-\infty}^{\infty} \frac{\chi(\omega_1)}{\omega_1 - \omega} d\omega_1 \tag{9.62}$$

$$\int_{(2)} = \int_B \frac{\chi(\omega_1)}{\omega_1 - \omega} d\omega_1 \tag{9.63}$$

$$\int_{(3)} = \int_C \frac{\chi(\omega_1)}{\omega_1 - \omega} d\omega_1 \tag{9.64}$$

where P (not polarization) represents the *Cauchy principal value*.

Assuming that $|\chi(\omega_1)| \to 0$ faster than $\left|\dfrac{1}{\omega_1}\right|$ as $\omega_1 \to \infty$, Eq. 9.63 becomes zero.

Also assume that

$$\omega_1 = \omega + \eta e^{i\theta} \tag{9.65}$$

$$\omega_1 - \omega = \eta e^{i\theta} \tag{9.66}$$

$$dω_1 = ηie^{iθ}dθ \qquad (9.67)$$

Substituting Eqs. 9.65–9.67 in Eq. 9.64, we get

$$\int_{(3)} = \int_π^0 \frac{χ(ω + ηe^{iθ})}{ηe^{iθ}}i\ ηe^{iθ}dθ \qquad (9.68)$$

$$\int_{(3)} = -iπχ(ω) \qquad (9.69)$$

Relations of Eqs. 9.61 and 9.69 let Eq. 9.62 to be

$$\int_{(1)} = P\int_{-∞}^∞ \frac{χ(ω_1)}{ω_1 - ω}dω = iπχ(ω) \qquad (9.70)$$

or $\quad χ(ω) = \dfrac{1}{iπ}P\int_{-∞}^∞ \dfrac{χ(ω_1)}{ω_1 - ω}dω \qquad (9.71)$

Now, separating Eq. 9.71 according to Eq. 9.60 into real and imaginary parts, we get

$$χ'(ω) = \frac{1}{π}P\int_{-∞}^∞ \frac{χ''(ω_1)}{ω_1 - ω}dω_1 \qquad (9.72)$$

$$χ''(ω) = -\frac{1}{π}P\int_{-∞}^∞ \frac{χ'(ω_1)}{ω_1 - ω}dω_1 \qquad (9.73)$$

We integrate over only the positive $ω_1$ by using the facts of Fourier transform that $χ'(ω)$ is even and $χ''(ω)$ is odd in $ω$. Multiplying the expressions in Eqs. 9.72 and 9.73 by $\dfrac{ω_1 + ω}{ω_1 + ω}$, we get

$$χ'(ω) = \frac{2}{π}P\int_0^∞ \frac{ω_1\ χ''(ω_1)}{ω_1^2 - ω^2}dω_1 \qquad (9.74)$$

$$χ''(ω) = -\frac{2}{π}P\int_0^∞ \frac{ω_1\ χ''(ω_1)}{ω_1^2 - ω^2}dω_1 \qquad (9.75)$$

Equations 9.74 and 9.75 are "one form of K–K relations," in the sense that the K–K relations have been widely applied, as in the following examples.

The K–K relation can be applied to the complex refractive index $ñ(ω)$. We identify a physical quantity $Q = Q_{real} + iQ_{imaginary}$ (response function of a system) as compared with the complex function $χ$ by

$$n(\omega) - 1 = \chi' \tag{9.76}$$

$$\kappa(\omega) = \chi'' \tag{9.77}$$

The real and imaginary components of the complex refractive index $\tilde{n}(\omega)$ can then be related as

$$n(\omega)-1=\frac{2}{\pi}P\int_0^\infty \frac{\omega_1\,\kappa(\omega_1)}{\omega_1^2-\omega^2}\,d\omega_1 = \frac{2}{\pi}P\int_0^\infty \frac{\alpha(\omega_1)}{\omega_1^2-\omega^2}\,d\omega_1 \tag{9.78}$$

$$\kappa(\omega)=-\frac{2}{\pi}P\int_0^\infty \frac{\omega\,n(\omega_1)}{\omega_1^2-\omega^2}\,d\omega_1 \tag{9.79}$$

In addition, the K–K relation can be applied to the reflection coefficient. If the normal reflection coefficient is related to the amplitude $r(\omega)$ and the phase angle $\theta(\omega)$ as

$$\tilde{r}_\perp = r(\omega)\,e^{i\theta} \tag{9.80}$$

and reflectance $\quad R=\left|\tilde{r}\right|^2 = r^2 \tag{9.81}$

we get

$$\ln\tilde{r}_\perp = 0.5\ln R(\omega) + i\theta(\omega) \tag{9.82}$$

We then identify

$$0.5\ln R(\omega) = \chi' \tag{9.83}$$

$$\theta(\omega) = \chi'' \tag{9.84}$$

Hence, the amplitude $r(\omega)$ and the phase angle $\theta(\omega)$ of the normal reflection coefficient are related as

$$\theta(\omega)=-\frac{2}{\pi}P\int_0^\infty \frac{\omega\,0.5\ln[R(\omega_1)]}{\omega_1^2-\omega^2}\,d\omega_1 \tag{9.85}$$

Thus, when R is experimentally measured, θ can be calculated. Once R and θ are known, the reflection coefficient $\tilde{r}$ can be calculated as well. Finally, the electrical conductivity can be analyzed.

On the other hand, *sum rules* hold between the imaginary part of the electric susceptibility $\chi''(\omega_1)$ and the absorption coefficient $\alpha(\omega_1)$ as a function of the plasma frequency ω_p of the electron gas in the material as

$$\int_{-\infty}^\infty \omega_1\chi''(\omega_1)\,d\omega_1 = \frac{\pi\omega_p^2}{2} \tag{9.86}$$

$$\int\limits_{-\infty}^{\infty} \alpha(\omega_1)\, d\omega_1 = \frac{\pi \omega_p^2}{2c} \tag{9.87}$$

Here the plasma frequency is

$$\omega_p = \sqrt{\frac{Ne^2}{m^* \varepsilon_0}} \tag{9.88}$$

where N and m^* denote the number of electrons within a unit volume and the effective mass of the electron involved in the optical transition, respectively. A relation holds between the real part of the electric susceptibility at zero frequency, $\chi'(0)$, and the imaginary part of the electric susceptibility at frequency ω_1, $\chi''(\omega_1)$:

$$\chi'(0) = \frac{2}{\pi} \int\limits_0^{\infty} \frac{\chi''(\omega_1)}{\omega_1}\, d\omega_1 \tag{9.89}$$

9.5 Problems

1. Prove that

$$\frac{\partial}{\partial t}\left(\vec{D} + \int \vec{J} dt\right) = \frac{\partial}{\partial t}\left(\varepsilon \vec{E} - \frac{1}{i\omega}\sigma \vec{E}\right)$$

2. Verify that the reflection coefficient $\tilde{r}_\alpha$ and the transmission coefficient $\tilde{t}_\alpha$ of α-polarized light at a plane boundary between medium 1 and medium 2 with dielectric constants $\tilde{\varepsilon}_1$ and $\tilde{\varepsilon}_2$, when light travels from medium 1 to medium 2, are

$$\frac{E_{r,s}}{E_{i,s}} = \tilde{r}_s = \tilde{t}_s - 1 = \frac{\cos\theta_1 - \sqrt{\dfrac{\tilde{\varepsilon}_2}{\varepsilon_1} - \sin^2\theta_1}}{\cos\theta_1 + \sqrt{\dfrac{\tilde{\varepsilon}_2}{\varepsilon_1} - \sin^2\theta_1}}$$

$$\frac{E_{r,p}}{E_{i,p}} = \tilde{r}_p = \tilde{t}_p - 1 = \frac{\dfrac{\tilde{\varepsilon}_2}{\varepsilon_1}\cos\theta_1 - \sqrt{\dfrac{\tilde{\varepsilon}_2}{\varepsilon_1} - \sin^2\theta_1}}{\dfrac{\tilde{\varepsilon}_2}{\varepsilon_1}\cos\theta_1 + \sqrt{\dfrac{\tilde{\varepsilon}_2}{\varepsilon_1} - \sin^2\theta_1}}$$

3. Verify that for a sufficiently large electrical conductivity σ, the skin depth is

$$\frac{c}{\omega\kappa} = \sqrt{\frac{2}{\sigma\omega\mu_0}}$$

4. Suppose that light is incident on a prism of refractive index n and of isosceles triangular shape with an apex angle β. When the angle of deviation δ is at the minimum δ_0, the optical path is symmetric as shown in the figure below. Use this fact to prove the relation

$$n = \frac{\sin\dfrac{\beta+\delta_0}{2}}{\sin\dfrac{\beta}{2}}$$

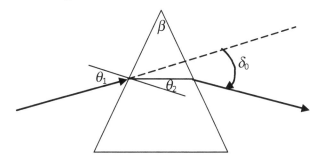

5. Prove that the power of an electromagnetic wave is absorbed when the phase of the vibration of the induced polarization is delayed compared to that of the electric field during its travel in a medium. This phenomenon is called dielectric loss. Let ϕ be the phase shift. Calculate the loss of the cycle-averaged power of the electromagnetic wave per unit time and unit volume.

6. Prove the relations

$$\chi'(\omega) = \frac{2}{\pi} P \int_0^\infty \frac{\omega_1\,\chi'(\omega_1)}{\omega_1^2 - \omega^2}\,d\omega_1$$

$$\chi''(\omega) = -\frac{2}{\pi} P \int_0^\infty \frac{\omega_1\,\chi'(\omega_1)}{\omega_1^2 - \omega^2}\,d\omega_1$$

7. Prove the relations

$$n(\omega)-1=\frac{2}{\pi}P\int_0^\infty \frac{\omega_1\,\kappa(\omega_1)}{\omega_1^2-\omega^2}\,d\omega_1 = \frac{2}{\pi}P\int_0^\infty \frac{\alpha(\omega_1)}{\omega_1^2-\omega^2}\,d\omega_1$$

$$\kappa(\omega)=-\frac{2}{\pi}P\int_0^\infty \frac{\omega\,n(\omega_1)}{\omega_1^2-\omega^2}\,d\omega_1$$

8. Prove the relations

$$\theta(\omega)=-\frac{2}{\pi}P\int_0^\infty \frac{\omega\,0.5\ln[R(\omega_1)]}{\omega_1^2-\omega^2}\,d\omega_1$$

9. Apply K–K relations in a complex dielectric function.
10. Apply K–K relations in Ohm's law.

Chapter 10

Classical Theory of Light–Matter Interaction II

10.1 Introduction

In this chapter, we will continue to understand the basic theorems related to the classical treatment of optical properties of matter. First, we will learn the Lorentz model of insulators (i.e., NaCl, NaF, NaBr), the Drude model of metals and semiconductors, and the theory of local field by Lorentz. We will then learn light emission from an electric dipole.

10.2 Lorentz Model

The Lorentz model describes most simply the dielectric responses of insulators, in which electrons are bound to positive ions (Fig. 10.1). The equation of motion of the electron in an oscillating electric field $\vec{E}$ with frequency ω will be

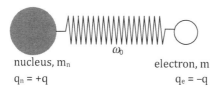

nucleus, m_n
$q_n = +q$

electron, m
$q_e = -q$

Figure 10.1 Electron bounded to the Bohr atomic model.

Optical Properties of Solids: An Introductory Textbook
Kitsakorn Locharoenrat
Copyright © 2016 Pan Stanford Publishing Pte. Ltd.
ISBN 978-981-4669-06-1 (Hardcover), 978-981-4669-07-8 (eBook)
www.panstanford.com

$$m\frac{d^2 X}{dt^2} = F_{\text{DRIVING}} + F_{\text{DAMPING}} + F_{\text{SPRING}} \tag{10.1}$$

$$m\frac{d^2 X}{dt^2} = F_{\text{DRIVING}} - m\Gamma_0 \frac{dX}{dt} - m\omega_0^2 X \tag{10.2}$$

$$m\left[\frac{d^2 X}{dt^2} + \Gamma_0 \frac{dX}{dt} + \omega_0^2 X\right] = qE = qE_0 e^{-i\omega t} \tag{10.3}$$

Here, m is the electron mass, X is the position of the electron relative to the atom, Γ_0 is the damping coefficient and also the bandwidth, ω_0 is the resonance frequency of an electron motion (a transverse wave), q is the absolute value of electron charge, ω is the angular frequency of the time-varying electric field, and E_0 is the amplitude of the electric field.

When we assume a solution to Eq. 10.3 as

$$X = X_0 e^{-i\omega t} \tag{10.4}$$

and substitute Eq. 10.4 in Eq. 10.3, the amplitude of the displacement X_0 is found as

$$m[-\omega^2 X_0 - i\omega\Gamma_0 X_0 + \omega_0^2 X_0]e^{-i\omega t} = qE_0 e^{-i\omega t} \tag{10.5}$$

$$X_0 = \frac{qE_0}{m(\omega_0^2 - \omega^2 - i\omega\Gamma_0)} \tag{10.6}$$

If the polarization $\vec{P}$ can be written as

$$\vec{P} = NqX = (\varepsilon - \varepsilon_0)\vec{E} \tag{10.7}$$

$$NqX_0 e^{-i\omega t} = (\varepsilon - \varepsilon_0)E_0 e^{-i\omega t} \tag{10.8}$$

and we substitute Eq. 10.6 in Eq. 10.8, we get

$$Nq\frac{qE_0}{m(\omega_0^2 - \omega^2 - i\omega\Gamma_0)}e^{-i\omega t} = (\varepsilon - \varepsilon_0)E_0 e^{-i\omega t} \tag{10.9}$$

Suppose we have N oscillators or electrons in a unit volume, and Nf_i of them have an Eigen frequency ω_j. The f_i is the oscillator strength. Thus, the dielectric function of this group of oscillators from Eq. 10.9 is

$$\varepsilon(\omega) = \varepsilon_0 + \frac{Nq^2}{m}\sum_j \frac{f_j}{(\omega_j^2 - \omega^2 - i\omega\Gamma_j)} \tag{10.10}$$

When we focus on one particular oscillator at $\omega = 0$, we define ε_l and ε_u as the dielectric functions at sufficiently lower and higher frequencies by

$$\varepsilon_0 = \varepsilon_u \tag{10.11}$$

$$\varepsilon(0) = \varepsilon_l \tag{10.12}$$

Substituting Eqs. 10.11 and 10.12 in Eq. 10.10, we get

$$\varepsilon_l = \varepsilon_u + \frac{Nq^2}{m\omega_0^2} \tag{10.13}$$

$$(\varepsilon_l - \varepsilon_u)\omega_0^2 = \frac{Nq^2}{m} \tag{10.14}$$

Substituting Eq. 10.14 in Eq. 10.10, we get

$$\varepsilon_l = \varepsilon_u + \frac{(\varepsilon_l - \varepsilon_u)\omega_0^2}{(\omega_0^2 - \omega^2 - i\omega\Gamma_0)} \tag{10.15}$$

$$\frac{\varepsilon_l}{\varepsilon_u} = 1 + \frac{\left(\dfrac{\varepsilon_l}{\varepsilon_u} - 1\right)\omega_0^2}{(\omega_0^2 - \omega^2 - i\omega\Gamma_0)} \tag{10.16}$$

When $\Gamma_0^2 \ll \omega_L^2 \approx \omega_0^2$, we introduce the Lyddane–Sachs–Teller relation as

$$\frac{\varepsilon_l}{\varepsilon_u} = \frac{\omega_L^2}{\omega_0^2} \tag{10.17}$$

Substituting Eq. 10.17 in Eq. 10.16, we have the dispersion relation as

$$\varepsilon_l = \varepsilon_u \left[1 + \frac{\omega_L^2 - \omega_0^2}{(\omega_0^2 - \omega^2 - i\omega\Gamma_0)} \right] \tag{10.18}$$

Where ω_L is the frequency of the longitudinal wave.

Equation 10.18 explains that in the frequency region of $\omega_0 - \Gamma_0 < \omega < \omega_L$, the material shows large optical absorption (A) and reflection (R), as seen in Fig. 10.2. In the remaining region, the material is transparent (T). In addition, when $\Gamma_0 = 0$ and in the frequency region of $\omega_0 < \omega < \omega_L$, the dielectric function will be real and negative ε', and thus the refractive index will be pure imaginary κ. Then a *total reflection* is observed.

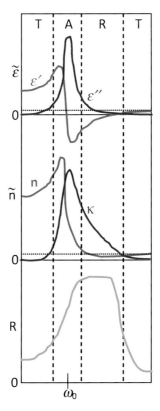

Figure 10.2 Dispersion relations $[\tilde{\varepsilon}(\omega)$ and $\tilde{n}(\omega)]$ and reflection (R) according to Lorentz model. A and T denote absorption and transparent, respectively.

10.3 Drude Model

"Electrons in metals (in conduction band as shown in Fig. 10.3) are loosely connected to their cores and can perform quasi-free movements," said Drude. This leads to a plasma-like behavior.

Figure 10.3 Plasma-like concept in metal.

We first consider the electric field $\vec{E} = E_0 e^{-i\omega t}$ to be incident on free electron gas, as shown in Fig. 10.1. Electrons then experience acceleration force $\left(m * \dfrac{d^2 X}{dt^2} \right)$ and electron collisions $\left(m * \tau^{-1} \dfrac{dX}{dt} \right)$. Suppose no restoring "spring" force is applied to the free electrons in metals and semiconductors as the electrons are not bound to the nuclei allowing these materials to conduct electricity, these conditions yield a natural frequency of $\omega_0 = 0$. Equation 10.3 then becomes

$$m * \left[\frac{d^2 X}{dt^2} + \tau^{-1} \frac{dX}{dt} \right] = qE = qE_0 e^{-i\omega t} \tag{10.19}$$

where τ is the average time interval between one scattering and another free electrons, and it is normally around 10^{-14} s. The $m*$ is the effective mass of an electron.

When we assume a solution to Eq. 10.19 of the form

$$X = X_0 e^{-i\omega t} \tag{10.20}$$

and substitute Eq. 10.20 in Eq. 10.19, X_0 is found as

$$m * [-\omega^2 X_0 - i\omega\tau^{-1} X_0] e^{-i\omega t} = qE_0 e^{-i\omega t} \tag{10.21}$$

$$X_0 = \frac{qE_0}{m * (-\omega^2 - i\omega\tau^{-1})} \tag{10.22}$$

From polarization (Eq. 10.8), we have

$$NqX_0 e^{-i\omega t} = (\varepsilon - \varepsilon_0) E_0 e^{-i\omega t}$$

Substituting Eq. 10.8 in Eq. 10.22, we find the dielectric function of this material as

$$\varepsilon = \varepsilon_0 - \frac{Ne^2}{m * \omega(\omega + i\tau^{-1})} \tag{10.23}$$

In the case of low frequency at $\omega \approx 0$, Eq. 10.23 becomes

$$\varepsilon = \varepsilon_0 + i\frac{Ne^2 \tau}{m * \omega} \tag{10.24}$$

If we define the electrical conductivity σ as

$$\sigma = \frac{Ne^2 \tau}{m *} \tag{10.25}$$

and substitute Eq. 10.25 in Eq. 10.24, we obtain

$$\varepsilon = \varepsilon_0 + i\frac{\sigma}{\omega} \tag{10.26}$$

In the case of high frequency at $\omega \gg \tau^{-1}$, Eq. 10.23 becomes

$$\varepsilon = \varepsilon_0 - \frac{Ne^2}{m^*\omega^2} \tag{10.27}$$

If we define the plasma frequency ω_p as

$$\omega_p = \sqrt{\frac{Ne^2}{m^*\varepsilon_0}} \tag{10.28}$$

and we substitute Eq. 10.28 in Eq. 10.27, we obtain

$$\varepsilon = \varepsilon_0 - \varepsilon_0 \frac{\omega_p^2}{\omega^2} \tag{10.29}$$

$$\varepsilon = \varepsilon_0 \left(1 - \frac{\omega_p^2}{\omega^2} \right) \tag{10.30}$$

Equation 10.30 reveals that as ω_p is plasma frequency, $\hbar\omega_p$ is located in visible or ultraviolet photon energy region as shown in Fig. 10.4.

Below this energy region, real metals behave like an ideal metal and their optical reflectivity tends to be close to unity. By contrast, beyond this energy, free carriers in the metals cannot follow a variation of the electric field and then the electromagnetic wave tends to penetrate and propagate freely into the metals. In addition, when $\omega = \omega_p$, $\varepsilon = 0$ according to Eq. 10.30, and the wavelength of light becomes infinite. This corresponds to a collective excitation called *plasmon* (Fig. 10.5).

Furthermore, at sufficiently high frequencies, the Drude model can also be applied to all materials. When ω become infinite, ε becomes ε_0 and the refractive index becomes unity. This is why we do not have good mirrors or lenses in X-ray or Gamma-ray regions. Figure 10.6 is an example of absorption-polarization analysis using the Drude model.

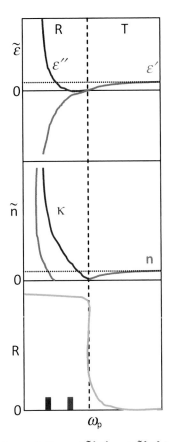

Figure 10.4 Dispersion relations $[\tilde{\varepsilon}(\omega)$ and $\tilde{n}(\omega)]$ and reflection (R) according to Drude model. T denotes transparent.

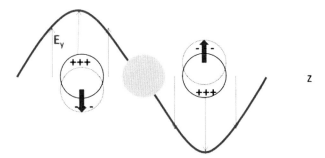

Figure 10.5 Electric oscillation in metal nanoparticles.

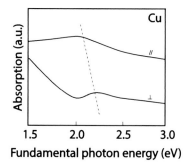

Figure 10.6 Absorption of Cu nanowires. $\perp$ (//) means the electric field is perpendicular (parallel) to the wire axes.

10.4 Theory of Local Field

The local electric field acting on atoms differs from a macroscopic electric field applied to a medium composed of these atoms. Suppose we have a void sphere instead of an atom and a macroscopic field $\vec{E}$ is applied to the whole medium, as shown in Fig. 10.7.

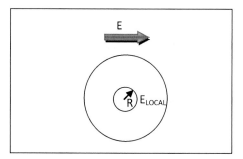

Figure 10.7 Lorentz local field.

Electric charges of amount $\vec{P} \cdot \vec{n}$ will appear on the internal face of the sphere. These charges will enhance the internal electric field at the center of the sphere, in addition to the average macroscopic field. Namely, we have

$$\vec{E}_{\text{LOCAL}} = \vec{E} + \frac{\vec{P}}{3\varepsilon_0} \tag{10.31}$$

This is called the *local electric field by Lorentz*. This model of the local electric field by Lorentz can be applied to the case of the local electrons of insulators. It is noted that in the case of the carriers traveling freely in the whole crystal, just as in the case treated by the Drude model, the average macroscopic field directly acts on the electrons. Figure 10.8 is an example of a local field effect using Eq. 10.31.

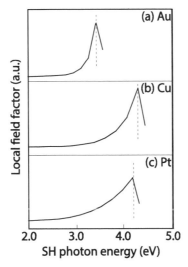

Figure 10.8 Local field enhancement factor versus SH photon energy for (a) Au, (b) Cu, and (c) Pt nanowires.

Next, the atomic or molecular electric dipole moment $\vec{M}$ is dependent on the local electric field strength at the atom or molecule as

$$\vec{M}_j = \alpha_j\,\vec{E}_{\mathrm{LOCAL}} \tag{10.32}$$

where α is the atomic or molecular polarizability.

We also define the material polarization as

$$\vec{P} = \sum_j N_j \vec{M}_j \tag{10.33}$$

where N is the concentration (number of atoms or molecules in a unit volume).

Substituting Eqs. 10.31 and 10.32 in Eq. 10.33, we get

$$\vec{P} = \sum_j N_j\,\alpha_j \vec{E}_{\mathrm{LOCAL}} \tag{10.34}$$

$$\vec{P} = \sum_j N_j \alpha_j \left(\vec{E} + \frac{\vec{P}}{3\varepsilon_0} \right) \tag{10.35}$$

where α is the polarizability.

We then apply

$$\vec{P} = (\varepsilon - \varepsilon_0)\vec{E} \tag{10.36}$$

and substitute Eq. 10.36 in Eq. 10.35 to finally obtain

$$(\varepsilon - \varepsilon_0)\vec{E} = \sum_j N_j \alpha_j \vec{E} + \frac{1}{3\varepsilon_0} \sum_j N_j \alpha_j (\varepsilon - \varepsilon_0)\vec{E} \tag{10.37}$$

$$(\varepsilon - \varepsilon_0) = \frac{3\varepsilon_0}{3\varepsilon_0} \sum_j N_j \alpha_j + \frac{\varepsilon}{3\varepsilon_0} \sum_j N_j \alpha_j - \frac{\varepsilon_0}{3\varepsilon_0} \sum_j N_j \alpha_j \tag{10.38}$$

$$(\varepsilon - \varepsilon_0) = \frac{2\varepsilon_0}{3\varepsilon_0} \sum_j N_j \alpha_j + \frac{\varepsilon}{3\varepsilon_0} \sum_j N_j \alpha_j \tag{10.39}$$

$$\frac{\varepsilon - \varepsilon_0}{\varepsilon + 2\varepsilon_0} = \frac{1}{3\varepsilon_0} \sum_j N_j \alpha_j \tag{10.40}$$

If we consider a transparent region and put

$$\varepsilon(\omega) = n^2(\omega) \tag{10.41}$$

in Eq. 10.40, we get the *Lorentz–Lorentz's relation*. On the other hand, when we put $\omega \rightarrow 0$ in Eq. 10.40, we get the *Clausius–Mossotti relation*.

10.5 Electromagnetic Field Radiation

The radiation field from current $\vec{i}_T$ is obtained from the vector potential $\vec{A}$ in Coulomb gauge ($\vec{i}_L = 0$) as

$$\vec{A}(x, y, z, t) = \frac{\mu_0}{4\pi} \iiint \frac{\vec{i}_T \left(x', y', z', \frac{tR}{c'} \right)}{R} dx' dy' dz' \tag{10.42}$$

in which $R \equiv |\vec{r} - \vec{r}'|$

Due to

$$\vec{E} = -\vec{\nabla}\phi - \frac{\partial \vec{A}}{\partial t} \tag{10.43}$$

and

$$\vec{B} = \vec{\nabla} \times \vec{A} = \frac{\vec{R} \times \vec{E}}{Rc'} \tag{10.44}$$

as well as when we expand the integrand $R = |\vec{r} - \vec{r}'|$ in Eq. 10.42 with respect to $\vec{r}'$, we obtain the multipole terms as

$$\vec{E}(\vec{r}, t) = -\frac{\mu_0}{4\pi r} \ddot{\vec{M}}_T\left(t - \frac{r}{c'}\right) + \frac{\mu_0 \vec{r}}{4\pi r^2 c'} \times \ddot{\vec{m}}\left(t - \frac{r}{c'}\right) - \frac{\mu_0 \vec{r}}{8\pi r^4 c'} \times$$

$$\left[\left\{Q(t - \frac{r}{c'})\vec{r}\right\} \times \vec{r}\right] + \cdots \tag{10.45}$$

where

$$\vec{M}(t) = \int \vec{r}' \rho(\vec{r}', t) d\vec{r}' \tag{10.46}$$

$$\vec{m}(t) = \frac{1}{2} \int \vec{r}' \times \vec{J}(\vec{r}', t) d\vec{r}' \tag{10.47}$$

$$Q_{ij}(t) = \left[\vec{r}_i' \vec{r}_j' - \frac{1}{3}\delta_{ij}\right] \rho(\vec{r}', t) d\vec{r}' \tag{10.48}$$

$\vec{M}, \vec{m}, Q$ are called the electric dipole moment, the magnetic dipole moment, and the electric quadrupole tensor, respectively. δ is the charge distribution. Fields radiated from these mutipoles are called the electric dipole field, the magnetic dipole field, and the electric quadrupole field, respectively. The magnetic dipole field and the electric quadrupole field are weaker than the electric dipole field by a factor of $(ak)^2$, where a is the size of an atom. The suffix T represents the transverse component with respect to the direction r.

For a point charge q moving with the velocity v, much slower than the velocity of light c', the electric dipole term in Eq. 10.45 becomes

$$\vec{E}(\vec{r}, t) = \frac{\mu_0}{4\pi R^3} \vec{R} \times (\vec{R} \times q\dot{\vec{v}}) \tag{10.49}$$

When a point charge q oscillates like $X = X_0 e^{-i\omega t}$ in the x-direction, Eq. 10.49 becomes

$$\vec{E}(\vec{r}, t) = \frac{\omega^2 \mu_0}{4\pi r^3} (\vec{r} \times qX_0 \vec{e}_x) \times \vec{r} e^{i(kr - \omega t)} \tag{10.50}$$

Then the Poynting vector $\vec{S}(\vec{r}, t)$ and its integration over an entire solid angle Ω are

$$\vec{S}(\vec{r},t) = \frac{\omega^4 \mu_0 \vec{r}}{16\pi^2 r^3 c'} q^2 X_0^2 \sin^2\theta \, \cos^2(kr - \omega t) \tag{10.51}$$

$$\int |\vec{S}| r^2 d\Omega = \frac{\omega^4 \mu_0}{6\pi c'} q^2 X_0^2 \cos^2(kr - \omega t) \tag{10.52}$$

If the charged, oscillating particle receives a friction force proportional to its velocity, the intensity spectrum of a radiated electromagnetic wave is

$$I(\omega) \propto |E(\omega)|^2 \propto \frac{1}{(\omega - \omega_0)^2 + (0.5\Gamma_0)^2} \tag{10.53}$$

Γ_0 is purely related to the reaction of the emission of the electromagnetic field by

$$\Gamma_0 = \frac{\mu_0 q^2 \omega_0^2}{6\pi m c'} \tag{10.54}$$

In addition, there are other factors that contribute to the friction forces, such as scattering with phonons. That is, when the electromagnetic field forces a bound electron to oscillate and thus creates an oscillating electric dipole with the same frequency as that of the incident field, then this dipole radiates a light field. This phenomenon is called *Rayleigh scattering*. From Eqs. 10.6 and 10.52, we can obtain the total power of the Rayleigh scattering as

$$W = \frac{\mu_0 q^2 \omega^4 |X_0|^2}{12\pi c'} \tag{10.55}$$

$$W = \frac{\mu_0 q^4 |E_0|^2}{12\pi m^2 c'} \cdot \frac{\omega^4}{(\omega^2 - \omega_0^2)^2 + \omega^2 \Gamma_0^2} \tag{10.56}$$

The cycle-averaged power of the incident electromagnetic field passing through a unit area is

$$\bar{\bar{S}} = \frac{|E_0|^2}{2\mu_0 c'} \tag{10.57}$$

The ratio of Eqs. 10.56 and 10.57 is defined as the scattering cross section (Fig. 10.9) given by

$$\sigma_R = \frac{W}{|\bar{\bar{S}}|} \tag{10.58}$$

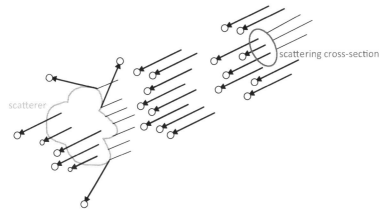

Figure 10.9 Scattering cross section is a probability that a photon beam impinging on the scatter is scattered into a unit solid angle in the given direction.

$$\sigma_R = \frac{\mu_0^2 e^4}{6\pi m^2} \frac{\omega^4}{(\omega^2 - \omega_0^2)^2 + \omega^2 \Gamma_0^2} \tag{10.59}$$

If we assume Γ_0, $\omega \ll \omega_0$, then we get

$$\sigma_R \approx \frac{\mu_0^2 e^4}{6\pi m^2} \frac{\omega^4}{\omega_0^4} \tag{10.60}$$

10.6 Problems

1. Verify that the dielectric function of N oscillators is

$$\varepsilon = \varepsilon_0 + \sum_j \frac{Nq^2 f_j}{m(\omega_j^2 - \omega^2 - i\omega\Gamma_j)}$$

2. Verify that the Lyddane–Sachs–Teller relation is

$$\frac{\varepsilon_l}{\varepsilon_u} = \frac{\omega_L^2}{\omega_0^2}$$

3. Explain the dispersion relation in a transparent region.
4. Explain the relation between the momentum scattering time τ and the damping constant Γ_0.
5. Prove that the Lorentz field is

$$\vec{E}_{\text{LOCAL}} = \vec{E} + \frac{\vec{P}}{3\varepsilon_0}$$

6. Prove the Lorentz–Lorentz's relation.

7. Prove the Clausius–Mossotti relation.

8. Prove that the radiation field is

$$\vec{E}(\vec{r},t) = -\frac{\mu_0}{4\pi r}\ddot{\vec{M}}_T\left(t - \frac{r}{c'}\right) + \frac{\mu_0 \vec{r}}{4\pi r^2 c'} \times \ddot{\vec{m}}\left(t - \frac{r}{c'}\right) - \frac{\mu_0 \vec{r}}{8\pi r^4 c'} \times$$

$$\left[\left\{Q(t - \frac{r}{c'})\vec{r}\right\} \times \vec{r}\right] + \cdots$$

9. Prove that the integration of the Poynting vector $\vec{S}(\vec{r},t)$ over a whole solid angle Ω is

$$\int \left|\vec{S}\right| r^2 d\Omega = \frac{\omega^4 \mu_0}{6\pi c'} q^2 X_0^2 \cos^2(kr - \omega t)$$

10. Prove that the optical reflectivity of a metal in vacuum at normal incidence is

$$R\perp = 1 - \sqrt{\frac{8\omega\varepsilon_0}{\sigma}}$$

for $\omega \ll \tau^{-1} \ll \omega_p$, where τ is the mean free time of the free carrier in a metal.

Chapter 11

Quantum Theory of Light–Matter Interaction

11.1   Introduction

In this chapter, we will learn the quantum mechanical treatment of optical properties of matter. In the previous chapter, we learned a number of theories within the classical framework. Now we will learn how they are treated in the quantum theory and see the distinction between quantum mechanical treatment and the classical one.

11.2   Quantum Theory of Matter

If a particle of mass m moves in a certain direction under a potential energy $V(r)$, we have the equation of motion of the system (called the *Schröedinger equation*) as

$$i\hbar\frac{\partial}{\partial t}\Psi(r,t) = -\frac{\hbar^2}{2m}\frac{\partial^2}{\partial r^2}\Psi(r,t) + V(r)\Psi(r,t) \tag{11.1}$$

in which the *quantum mechanical wave function* is

$$\Psi(\vec{r}_1,\vec{r}_2,...\vec{r}_N,t) \tag{11.2}$$

The wave function represents the profile of the deformed matter (medium), as shown in Fig. 11.1.

Optical Properties of Solids: An Introductory Textbook
Kitsakorn Locharoenrat
Copyright © 2016 Pan Stanford Publishing Pte. Ltd.
ISBN 978-981-4669-06-1 (Hardcover), 978-981-4669-07-8 (eBook)
www.panstanford.com

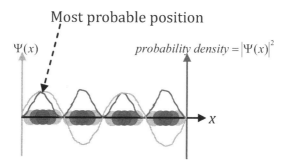

Figure 11.1   Quantum wave function.

The *probability density* at which particles $1 \rightarrow N$ are found in a small volume fraction $dr_1 \rightarrow dr_N$ near the point $r_1 \rightarrow r_N$ is

$$\left|\Psi(\vec{r}_1,\vec{r}_2,...\vec{r}_N,t)\right|^2 d\vec{r}_1 d\vec{r}_2...d\vec{r}_N \tag{11.3}$$

with normalization condition as

$$\int \Psi^*(\vec{R},t)\Psi(\vec{R},t)\,d\vec{R} = \int \left|\Psi(\vec{R},t)\right|^2 d\vec{R} = 1 \tag{11.4}$$

If the Hamiltonian operator is defined as

$$\widehat{H} = -\frac{\hbar^2}{2m}\frac{\partial^2}{\partial r^2} + V(r) \tag{11.5}$$

and we substitute Eq. 11.5 in Eq. 11.1, the Schröedinger equation can be rewritten as

$$\widehat{H}\Psi(\vec{R},t) = i\hbar\frac{\partial}{\partial t}\Psi(\vec{R},t) \tag{11.6}$$

If the wave function can be defined as

$$\Psi(\vec{R},t) = \psi(\vec{R})T(t) \tag{11.7}$$

and we substitute Eq. 11.7 in Eq. 11.6, we get

$$\widehat{H}[\psi(\vec{R})T(t)] = i\hbar\frac{\partial}{\partial t}[\psi(\vec{R})T(t)] \tag{11.8}$$

$$T(t)\widehat{H}\psi(\vec{R}) = i\hbar\psi(\vec{R})\frac{\partial}{\partial t}T(t) \tag{11.9}$$

$$\frac{1}{\psi(\vec{R})}\widehat{H}\psi(\vec{R}) = \frac{i\hbar}{T(t)}\frac{\partial}{\partial t}T(t) = W \tag{11.10}$$

From Eq. 11.10, the time-dependent Schröedinger equation is

$$\frac{i\hbar}{T(t)}\frac{\partial}{\partial t}T(t) = W \qquad (11.11)$$

$$\frac{\partial}{\partial t}T(t) = \frac{W}{i\hbar}T(t) \qquad (11.12)$$

with a solution

$$T(t) = e^{-\frac{iWt}{\hbar}} \qquad (11.13)$$

Also from Eq. 11.10, a time-independent Schröedinger equation is

$$\frac{1}{\psi(\vec{R})}\hat{H}\psi(\vec{R}) = W \qquad (11.14)$$

$$\hat{H}\psi(\vec{R}) = W\psi(\vec{R}) \qquad (11.15)$$

This system is the general case of a stationary state. This equation is also called the eigenvalue equation in which W is the eigenvalue of the system (constant) and ψ is the eigenfunction (wavefunction).

If a possible value of observable $\hat{A}$ is defined as

$$\langle A \rangle = \int \Psi^*(\vec{R},t)\,\hat{A}\,\Psi(\vec{R},t)\,d\vec{R} = \left\langle \Psi^*(\vec{R},t)\middle|\hat{A}\middle|\Psi(\vec{R},t)\right\rangle \qquad (11.16)$$

substituting Eq. 11.16 in Eq. 11.15, we get

$$\hat{A}\Psi_{\text{System}}(\vec{R},t) = \sum \langle A \rangle\, \psi(\vec{R})e^{-\frac{iWt}{\hbar}} \qquad (11.17)$$

Next, if the Hamiltonian operator is separated into unperturbed and perturbation parts as

$$\hat{H} = \hat{H}_0 + \hat{H}' \qquad (11.18)$$

we will finally have a solution to the energy W and the wavefunction ψ as

$$\tilde{W}_m = W_m + \left\langle m\middle|\hat{H}'\middle|m\right\rangle + \sum_{k \neq m}\frac{\left|\left\langle k\middle|\hat{H}'\middle|m\right\rangle\right|^2}{W_m - W_k} + \cdots \qquad (11.19)$$

$$\tilde{\psi}_m = \middle|m\right\rangle + \sum_{k \neq m}\middle|k\right\rangle\frac{\left\langle k\middle|\hat{H}'\middle|m\right\rangle}{W_m - W_k} + \cdots \qquad (11.20)$$

11.3 Semi-Classical Quantum Mechanical Treatment

Here we will determine the dielectric response of matter under the light frequency by solving a time-dependent perturbation theory (Fig. 11.2).

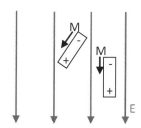

Figure 11.2 Dielectric response of matter with a light.

The perturbation Hamiltonian is

$$\widehat{H}' = -\overrightarrow{M} \cdot \vec{E}_0 \cos(\omega t) \tag{11.21}$$

$$\widehat{H}' = -\widehat{\mu}\, \vec{E}_0 \cos(\omega t) \tag{11.22}$$

where $\overrightarrow{M} = -\sum_j e\vec{r}_j$ is the atomic dipole moment and $\widehat{\mu}$ is its projection onto the direction of electric field. Let the wave function be

$$\Psi_{\text{System}}(\vec{R}, t) = \sum_n b_n(t)\Psi_n(\vec{R}, t) \tag{11.23}$$

$$\Psi_{\text{System}}(\vec{R}, t) = \sum_n b_n(t)\psi(\vec{R})e^{-\frac{iWt}{\hbar}} \tag{11.24}$$

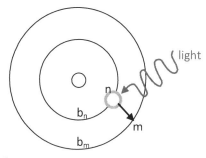

Figure 11.3 An electronic transition from *n* state to *m* state.

As shown in Fig. 11.3, if a population at level n is proportional to the transition probability $|b_n(t)|^2$, we have

$$i\hbar \frac{\partial b_m(t)}{\partial t} = \sum_n b_n(t) H'_{mn} e^{i\omega_{mn}t}$$

(11.25)

where $H'_{mn} = \langle m|\hat{H}'|n \rangle$ and $\omega_{mn} = \dfrac{(W_m - W_n)}{\hbar}$

Now we consider a two-level system ($|1\rangle, |2\rangle$), as illustrated in Fig. 11.4, and put $\omega_0 = \omega_{21}$ and $\Gamma_0 = \tau_2^{-1}$, where the latter is the life time of the electronic level 2.

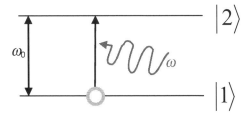

Figure 11.4   A two-electronic level from level 1 to level 2.

From Eqs. 11.22 and 11.25, we get

$$\frac{\partial b_2(t)}{\partial t} = \frac{i\mu_{21}E_0}{\hbar} \cos(\omega t) e^{(i\omega_0 t)} b_1(t) - \frac{\Gamma_0}{2} b_2(t)$$

(11.26)

In a room temperature environment ($k_B T \ll \hbar\omega_0$), we assume that $b_1(t) = 1$. Equation 11.26 then becomes

$$b_2(t) = \frac{\mu_{21}E_0}{2\hbar} \left[\frac{e^{i(\omega_0 + \omega)t}}{\omega_0 + \omega - \dfrac{i\Gamma_0}{2}} + \frac{e^{i(\omega_0 - \omega)t}}{\omega_0 - \omega - \dfrac{i\Gamma_0}{2}} \right]$$

(11.27)

The material polarization is

$$P(t) = N_0 < \hat{\mu} >$$

(11.28)

$$P(t) = N_0 < \Psi |\hat{\mu}| \Psi >$$

(11.29)

$$P(t) = N_0 \int (b_1^* \Psi_1^* \hat{\mu} \Psi_2 b_2 + b_2^* \Psi_2^* \hat{\mu} \Psi_1 b_1) \, d\vec{R}$$

(11.30)

If we substitute Eq. 11.27 in Eq. 11.30, we get

$$P(t) = \frac{N_0 |\mu_{21}|^2 E_0}{2\hbar} \cdot \left[\frac{e^{i\omega t}}{\omega_0 + \omega - \dfrac{i\Gamma_0}{2}} + \frac{e^{-i\omega t}}{\omega_0 - \omega - \dfrac{i\Gamma_0}{2}} + C.C. \right] \quad (11.31)$$

Due to

$$P(t) = \frac{\varepsilon_0 E_0}{2} \left[\chi(\omega) e^{-i\omega t} + \chi(-\omega) e^{i\omega t} \right] \quad (11.32)$$

if we put Eq. 11.32 in Eq. 11.31, we get

$$\chi(\pm\omega) = \frac{N_0 |\mu_{21}|^2}{\varepsilon_0 \hbar} \cdot \frac{2\omega_0}{\omega_0^2 - \omega^2 + \left(\dfrac{\Gamma_0}{2}\right)^2 \mp i\omega\Gamma_0} \quad (11.33)$$

When we have several resonant levels, the electric susceptibility is

$$\chi(\pm\omega) = \frac{N}{3\varepsilon_0 \hbar} \sum_j \frac{2\omega_{jg} |\vec{M}_{jg}|^2}{\omega_{jg}^2 - \omega^2 + \left(\dfrac{\Gamma_j}{2}\right)^2 \mp i\omega\Gamma_j} \quad (11.34)$$

If we define f_{jg} as the oscillator strength by

$$f_{jg} = \frac{2m\omega_{jg} |\vec{M}_{jg}|^2}{3\hbar e^2} \quad (11.35)$$

and it fulfills the relations in one-electron or multi-electron systems by

$$\sum_j f_{jg} = 1 \quad \text{or} \quad \sum_j f_{jg} = Z \quad (11.36)$$

substituting Eq. 11.35 in Eq. 11.34, we finally get

$$\chi(\pm\omega) = \frac{Ne^2}{\varepsilon_0 m} \sum_j \frac{f_{jg}}{\omega_{jg}^2 - \omega^2 + \left(\dfrac{\Gamma_j}{2}\right)^2 \mp i\omega\Gamma_j} \quad (11.37)$$

Let us think of a transition from n to m. We next calculate the intensity of the second term of Eq. 11.27 as

$$|b_m(t)|^2 = \frac{|\vec{M}_{mn} \cdot \vec{E}_0|^2}{\hbar^2} \cdot \frac{\sin^2\left\{ \dfrac{1}{2}(\omega_{mn} - \omega)t \right\}}{(\omega_{mn} - \omega)^2} \quad (11.38)$$

$$|b_m(t)|^2 = \frac{\pi |\vec{M}_{mn} \cdot \vec{E}_0|^2}{2\hbar^2} t\delta(\omega_{mn} - \omega) \tag{11.39}$$

This relation is consistent with the quantum mechanical uncertainty relation between time and energy. From Eq. 11.39, we can have a transition probability per unit time as (m and n states are not degenerate)

$$w_{mn} = \frac{\pi |\vec{M}_{mn} \cdot \vec{E}_0|^2}{2\hbar^2} \delta(\omega_{mn} - \omega) \tag{11.40}$$

If the spectrum has a shape of $f(\omega)$, the state m has a degeneracy g_m, and the photon density is $\rho_\omega = n\hbar\omega$, we have

$$w_{mn} = B_{mn} f(\omega) \rho_\omega \tag{11.41}$$

$$B_{mn} = \frac{g_m \pi}{3\varepsilon_0 \hbar^2} |\vec{M}_{mn}|^2 \tag{11.42}$$

where B_{mn} is defined as *Einstein's B coefficient*.

Putting F as photon flux, σ defined in $W = F\sigma$ as the transition (absorption) cross section, and N_g as the atom density at ground state, we have the absorption coefficient $\alpha(\omega)$ as

$$\alpha(\omega) = \frac{\hbar\omega}{c} B_{mg} f(\omega) N_g \tag{11.43}$$

$$\alpha(\omega) = \sigma_{mg}(\omega) N_g \tag{11.44}$$

11.4 Fully Quantum Mechanical Treatment

The Hamiltonian of the charged particles under the electromagnetic field can be obtained using Direc's substitution of

$$\vec{p} = \vec{p} + e\vec{A} \tag{11.45}$$

in the Hamiltonian of free electrons as

$$\frac{\vec{p} \cdot \vec{p}}{2m} = \frac{1}{2m}(\vec{p} + e\vec{A})^2 \tag{11.46}$$

Expanding this final expression, we have the perturbation Hamiltonian. In Coulomb gauge, the perturbation Hamiltonian is

$$\widehat{H} = \frac{e}{m}\widehat{p} \cdot \widehat{A} \tag{11.47}$$

By using the vector potential

$$\vec{A}_\lambda = \sqrt{\frac{\hbar}{2\varepsilon_0 V \omega_\lambda}}\vec{e}_\lambda \left[\widehat{a}_\lambda e^{i(\vec{k}_\lambda \cdot \vec{r} - \omega_\lambda t)} + \widehat{a}_\lambda^+ e^{-i(\vec{k}_\lambda \cdot \vec{r} - \omega_\lambda t)}\right] \tag{11.48}$$

and substituting Eq. 11.48 in Eq. 11.47, we have

$$\widehat{H} = \frac{e}{m}\sum_\lambda \sqrt{\frac{\hbar}{2\varepsilon_0 V \omega_\lambda}}\widehat{p} \cdot \vec{e}_\lambda \left[\widehat{a}_\lambda e^{i(\vec{k}_\lambda \cdot \vec{r} - \omega_\lambda t)} + \widehat{a}_\lambda^+ e^{-i(\vec{k}_\lambda \cdot \vec{r} - \omega_\lambda t)}\right] \tag{11.49}$$

Putting the exponent as zero, as is known as the dipole approximation, we have the perturbation Hamiltonian as

$$\widehat{H}_{E1} = \frac{e}{m}\sum_\lambda \sqrt{\frac{\hbar}{2\varepsilon_0 V \omega_\lambda}}\widehat{p} \cdot \vec{e}_\lambda \left[\widehat{a}_\lambda e^{-i\omega_\lambda t} + \widehat{a}_\lambda^+ e^{i\omega_\lambda t}\right] \tag{11.50}$$

If we define a matrix component of the momentum operator as

$$< f \left|\widehat{p}\right| g > = \vec{p}_{fg} = im\omega_{fg}\vec{r}_{fg} \tag{11.51}$$

and we multiply this by elementary charge, we have the dipole moment operator. In addition, if we expand the exponent part of Eq. 11.49, we have the multi-pole expansion. Then the Hamiltonian of the interaction of light field together with the material is

$$H' = -\vec{M} \cdot \vec{E}(0,t) - \vec{m} \cdot \vec{B}(0,t) - \frac{1}{2}(Q\vec{\nabla}) \cdot \vec{E}(0,t) + \cdots \tag{11.52}$$

Next we consider the generalized form of Eq. 11.40. Let the Hamiltonian be

$$\widehat{H} = \widehat{H}_0 + \widehat{H}_R + \widehat{H}'(t) \tag{11.53}$$

H_0 and H_R denote the unperturbed Hamiltonians of the atomic system and the radiation field, respectively. $\widehat{H}'$ denotes the perturbed Hamiltonian of the interaction between the atomic system and the radiation field. As a generalized form of Eq. 11.40, we have the transition probability per unit time as

$$w_{mn} = \frac{2\pi}{\hbar}\left|H'_{MI}\right|^2 \delta(W_M - W_I) \tag{11.54}$$

where $\quad H'_{MI} \equiv \left\langle M \left| e^{-i\frac{\widehat{H}_R}{\hbar}t}\widehat{H}'(t)e^{i\frac{\widehat{H}_R}{\hbar}t} \right| I \right\rangle \tag{11.55}$

in which I and M are initial and intermediate states, respectively. This relation can be named *Fermi's golden rule*. The transition probability of the second-order optical process, such as optical second harmonic generation or Raman scattering, is obtained through the second-order perturbation theory as

$$w_{KI}^{(2)} = \frac{2\pi}{\hbar} \left| \sum_M \frac{H'_{KM} H'_{MI}}{\hbar \omega_{IM}} \right| \delta(W_M - W_I) \tag{11.56}$$

In general, the transition probability of *spontaneous photon emission* of light, as shown in Fig. 11.5, is obtained quantum mechanically, and not semi-classically, using Fermi's golden rule as follows:

$$w_{GM} = \sum_\lambda \frac{2\pi}{\hbar} \left| \left\langle g, n_\lambda + 1 \left| \hat{H}'_\lambda \right| m, n_\lambda \right\rangle \right|^2 \delta(\hbar \omega_{mg} - \hbar \omega_\lambda) \tag{11.57}$$

$$w_{GM} = \sum_\lambda \frac{\pi \omega_\lambda}{\hbar \varepsilon_0 V} \left| \vec{M}_{gm} \cdot \vec{e}_\lambda \right|^2 (n_\lambda + 1) \, \delta(\omega_{mg} - \omega_\lambda) \tag{11.58}$$

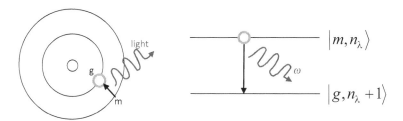

Figure 11.5 Electronic level from level *m* to level *g*.

Equation 11.58 consists of a part proportional to the number of photons and a part independent of the number of photons. The probability of *induced photon emission* of light is equal to that of *induced photon absorption* if the initial state and the final state have no degeneracy. Spontaneous emission can be explained only by a theory, including quantization of the radiation field. It is an induced emission of light triggered by a zero-point vibration of the radiation field. The transition probability of spontaneous emission process is

$$A_{gm} = \left| \vec{M}_{gm} \right|^2 \frac{\omega_{mg}^3}{3\pi \varepsilon_0 \hbar c^3} \tag{11.59}$$

where A_{gm} is defined as *Einstein's A coefficient*.

On the other hand, the transition probability of the Raman scattering process is

$$w_{fg} = \frac{n_1 \omega_1 \omega_2^3}{2\pi\hbar^2 \varepsilon_0^2 V c^3} \left| \sum_m \left[\frac{(\vec{M}_{fm} \cdot \vec{e}_2)(\vec{M}_{mg} \cdot \vec{e}_1)}{\omega_{mg} - \omega_1} + \frac{(\vec{M}_{fm} \cdot \vec{e}_1)(\vec{M}_{mg} \cdot \vec{e}_2)}{\omega_{mg} + \omega_2} \right] \right|^2$$

(11.60)

Finally, if the quantum state changes into another one, we call this phenomenon damping. The damping constant of an energy level is proportional to the speed of its damping by definition. If an atomic electron system interacts with the radiation field, a level shift and broadening occur.

$$\gamma = w_M + i\Delta_M$$ (11.61)

Here, Δ_M is called *Lamb shift* and represents the energetic shift of an electronic level as a result of its reaction to the force coming from the radiation field. Due to this effect, the energy of the 2s state of a hydrogen atom is higher than that of the sp state by 1057.8 MHz, and the energy level of the 1s state of a hydrogen atom is higher by 8.2 GHz than that calculated with no interaction with the radiation field considered.

11.5 Problems

1. Prove that the time-dependent Schröedinger equation is

$$i\hbar \frac{\partial}{\partial t} \Psi(r,t) = -\frac{\hbar^2}{2m} \frac{\partial^2}{\partial r^2} \Psi(r,t) + V(r)\Psi(r,t)$$

2. If the Hamiltonian operator is separated into unperturbed and perturbation parts as $\hat{H} = \hat{H}_0 + \hat{H}'$, determine a solution in terms of the wave function ψ and energy W.

3. If a population at level n is proportional to the transition probability $|b_n(t)|^2$, show that

$$i\hbar \frac{\partial b_m(t)}{\partial t} = \sum_n b_n(t) H'_{mn} e^{i\omega_{mn}t}$$

4. Prove that the transition probability is

$$|b_m(t)|^2 = \frac{|\vec{M}_{mn} \cdot \vec{E}_0|^2}{\hbar^2} \cdot \frac{\sin^2\left\{\frac{1}{2}(\omega_{mn} - \omega)t\right\}}{(\omega_{mn} - \omega)^2}$$

5. When F is a function of mechanical variables, the Heisenberg equation of motion for F is

$$\frac{dF_{jg}}{dt} = \left\langle j \left| \frac{\partial \hat{F}}{\partial t} \right| g \right\rangle + \frac{1}{i\hbar} \langle j | \hat{F}\hat{H} - \hat{H}\hat{F} | g \rangle$$

   Suppose that $|g\rangle$ and $|j\rangle$ are eigenfunctions of $\hat{H}$. Calculate the possible momentum matrix element of a particle.

6. Under a dipole approximation, prove that the perturbation Hamiltonian is

$$\hat{H}_{E1} = \frac{e}{m} \sum_\lambda \sqrt{\frac{\hbar}{2\varepsilon_0 V \omega_\lambda}} \, \hat{p} \cdot \hat{e}_\lambda \left[\hat{a}_\lambda e^{-i\omega_\lambda t} + \hat{a}_\lambda^+ e^{i\omega_\lambda t} \right]$$

7. Explain Fermi's golden rule related to fully quantum mechanical treatment of light interacted with matter.

8. Prove that the transition probability of photon emission is

$$w_{GM} = \sum_\lambda \frac{\pi \omega_\lambda}{\hbar \varepsilon_0 V} \left| \vec{M}_{gm} \cdot \vec{e}_\lambda \right|^2 (n_\lambda + 1) \, \delta(\omega_{mg} - \omega_\lambda)$$

9. Calculate Einstein's B coefficient from

$$w_{GM} = \sum_\lambda \frac{\pi \omega_\lambda}{\hbar \varepsilon_0 V} \left| \vec{M}_{gm} \cdot \vec{e}_\lambda \right|^2 (n_\lambda + 1) \, \delta(\omega_{mg} - \omega_\lambda)$$

   and also show that it is consistent with

$$B_{mn} = \frac{g_m \pi}{3\varepsilon_0 \hbar^2} \left| \vec{M}_{mn} \right|^2.$$

10. Suppose that a two-level electronic system with Boltzmann distribution is in a thermal equilibrium with a radiation field with Planck's distribution of temperature T. Prove that the probabilities of the emission and the absorption of electromagnetic waves by the two-level electronic system are the same.

Chapter 12

Electron–Nuclei Interaction

12.1 Introduction

In this chapter, we will learn the electronic states in interaction with molecular vibrations or phonons in condensed matter. Especially, we will study the optical spectra from localized electronic states and the elements determining the spectral shapes.

12.2 Separation of Motions of Electrons and Nuclei

The speed of motion of a nuclei is quite different from that of electrons. Hence, frequencies of their motions are quite different. The ratio of the frequencies of the motion of electrons to those of molecular vibration and molecular rotation is $10^4{:}10^2{:}1$. In such a case, we can consider their motions separately. For instance, suppose a lot of bees are flying around a lion, as shown in Fig. 12.1. The motions of the bees are so quick that the bees feel that the lion is not moving.

The lion is indeed annoyed at the bees flying around it, but it is not interested in the detailed position of the bees' flight path. Here it may be all right if we assume that the lion is not moving when we

Optical Properties of Solids: An Introductory Textbook
Kitsakorn Locharoenrat
Copyright © 2016 Pan Stanford Publishing Pte. Ltd.
ISBN 978-981-4669-06-1 (Hardcover), 978-981-4669-07-8 (eBook)
www.panstanford.com

deal with the motion of the bees, and if we consider only the average distribution density of the bees when we deal with the motion of the lion. In this way, we can separately deal with the motions of the bees and the lion. This does not mean that the motion of the bees has no effect on the motion of the lion. The average density distribution of the bees determined by their motions indeed influences the emotion and thus the motion of the lion. For example, the lion may be nervous enough to start moving. This effect of average distribution of light particles on heavier particles is seen in the electron–lattice (e-L) interaction in molecules and solid-state materials. It can be explained by solving Schröedinger equations with a Hamiltonian $\hat{H}$ by

$$\hat{H}\psi = W\psi \tag{12.1}$$

This equation describes the motions of the electrons bound in a molecule and of the nuclei, and we can separate the wave functions as

$$\psi_{\text{molecule}} = \psi_{\text{electrons}}\,\psi_{\text{nuclei}} \tag{12.2}$$

This equation can also be called the *Born–Oppenheimer approximation* or *adiabatic approximation.*

Figure 12.1 Model to explain motions of electrons and nuclei.

12.3 Lattice Vibrations

The interaction potential within a molecule can be regarded as that of an ideal spring if the displacement of the interatomic distance is small. In Fig. 12.2, suppose two atoms of weights M_a and M_b and of positions x_a and x_b are connected with a spring of spring constant k.

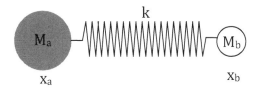

Figure 12.2   Interaction potential within a molecule.

Let us also suppose that these atoms move only on the x-axis and that the equilibrium distance between the two atoms is a. The equation of motion of the two atoms is

$$M_a\ddot{x}_a = k(x_b - x_a - a) \tag{12.3}$$

$$M_b\ddot{x}_b = -k(x_b - x_a - a) \tag{12.4}$$

If Eq. 12.3 is subtracted from Eq. 12.4, we get

$$\ddot{x}_a - \ddot{x}_b = k\left(\frac{1}{M_a} + \frac{1}{M_b}\right)(x_b - x_a - a) \tag{12.5}$$

If a reduced mass μ is defined as

$$\frac{1}{\mu} = \frac{1}{M_a} + \frac{1}{M_b} \tag{12.6}$$

and q is the distance between the two atoms, substituting Eq. 12.6 in Eq. 12.5, we get

$$\mu\ddot{q} = -kq \tag{12.7}$$

Equation 12.7 tells us that the center of mass of this molecule travels with a constant velocity and that the relative motion of the atoms is harmonic oscillation. Also called an internal coordinate, q describes the internal motion of the molecule. When q describes a harmonic oscillator after the transformation of the coordination system, as in the above case, q is also called a normal coordinate. A normal coordinate is the basis of the eigenenergy of the equation of the vibrational motion. The eigenfrequency ω_v and the quantum mechanical eigenenergy W_n of the harmonic oscillation described by Eq. 12.7 are

$$\omega_v = \sqrt{\frac{k}{\mu}} \tag{12.8}$$

$$W_n = \hbar\omega_v\left(n+\frac{1}{2}\right) \tag{12.9}$$

A similar frame of discussion can be applied to a system of general numbers of atoms. In crystals, we treat lattice vibration in the same way as we treat the vibration of molecules, except that a lot of vibrating atoms make the mode density high and the vibration travels as a wave. When a three-dimensional unit cell of a crystal contains more than two atoms, there exist longitudinal (L) and transverse (T) modes as well as optical (O) and acoustic (A) modes.

As seen in Fig. 12.3, an acoustic mode is a sound wave in which adjacent atoms vibrate in the same phase, whereas an optical mode is a local mode in which adjacent atoms vibrate in opposite phases. Combining these modes, we can obtain TO, TA, LO, and LA modes.

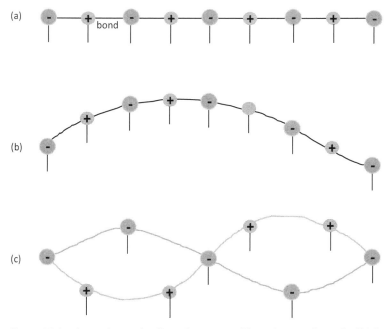

Figure 12.3 Acoustic mode (low frequency, b) and optical mode (high frequency, c) with respect to one-dimensional linear chain (a).

12.4 Absorption Spectrum

The values of the normal coordinates Q at the potential minimum of ground state U_g and excited state U_e are naturally different from each other, because the spring constants of the springs made from the ground- and excited-state electronic wavefunctions are different. For this reason, we insert the term $-cQ_1$ in Eq. 12.13, which has the effect of shifting the equilibrium interatomic distance of the molecules in question. To discuss the optical transition under the Born–Oppenheimer approximation mentioned in the previous section, we first consider a molecular system embedded in a solid-state matrix as a typical example (Fig. 12.4).

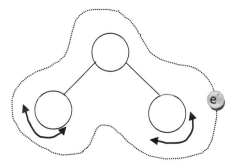

Figure 12.4   Three nuclei embedded in a molecular system surrounded by electron clouds.

On one hand, in the case of high temperature, the electronic transition occurs vertically in the potential energy diagram independently of the temperature according to the *Franck–Condon principle*, as seen in Fig. 12.5.

If the population distribution $P_g(Q_1)$ as a function of Q_1 is defined as

$$P_g(Q_1) = \sqrt{2\pi k_B T}\, e^{-\frac{Q_1^2}{2k_B T}} \tag{12.10}$$

the absorption $A(\hbar\omega)$ can be written as

$$A(\hbar\omega) \propto \int P_g(Q_1)\, \delta(U_e - U_g - \hbar\omega)\, dQ_1 \tag{12.11}$$

Using the relations as

$$U_g(Q_1) = \frac{1}{2}Q_1^2 \tag{12.12}$$

$$U_e(Q_1) = W_0 - cQ_1 + \frac{1}{2}Q_1^2 \tag{12.13}$$

$$W_{\text{Lattice relaxation}} = \frac{1}{2}c^2 \tag{12.14}$$

and substituting them in Eq. 12.11, we get the Condon approximation as

$$A(\hbar\omega) \propto e^{-\frac{(W_0 - \hbar\omega)^2}{4k_B T W_{\text{LR}}}} \tag{12.15}$$

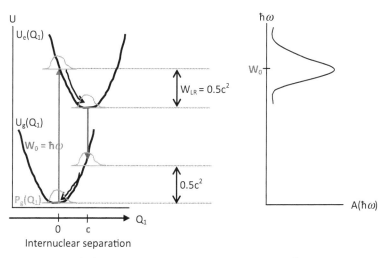

Figure 12.5   Adiabatic potential energy of the form $0.5kx^2$ as a function of the interaction mode. Subscripts g and e denote the ground and the excited electronic states, respectively.

Equation 12.15 is a Gaussian with peak at W_0 and a width proportional to $T^{0.5}$. The peak energy W_0 is equal to the average value of $U_e\text{-}U_g$, and the energy width comes from the fluctuation of $U_e\text{-}U_g$.

On the other hand, in the case of an intermediate temperature, vibrational structures within the electronic transition appear in the absorption spectrum, and the optical transition will then be in a quantized potential system, as shown in Fig. 12.6. The normal coordinate in the horizontal axis is generalized into a configuration coordinate representing atomic distances, bond angles, and general displacements of nuclei.

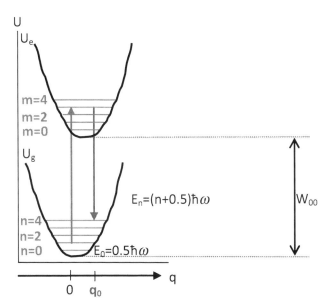

Figure 12.6 Adiabatic potential U curves for the ground and the excited vibrational states. $q = 0$ is the equilibrium position between the nuclei.

The absorption $A(\hbar\omega)$ can be written as

$$A(\hbar\omega) \propto \sum_n \sum_m \rho_{gn} \left| \langle \Psi_{em} | \hat{\mu} | \Psi_{gn} \rangle \right|^2 \delta(W_{em} - W_{gn} - \hbar\omega) \quad (12.16)$$

$$\rho_{gn} = \frac{e^{-\frac{W_{gn}}{k_B T}}}{\sum_n e^{-\frac{W_{gn}}{k_B T}}} \quad (12.17)$$

If we assume that the Born–Oppenheimer approximation and the Condon approximation (dipole moment μ_{eg} is independent of q) are valid, the absorption spectrum A at $T = 0$ from Eq. 12.17 is modified as

$$A(\hbar\omega) \propto \sum_m F_{m0} \, \delta(\hbar\omega - W_{00} - m\hbar\omega_v) \quad (12.18)$$

where the Frank–Condon factor is

$$F_{mn} = \left| \int \phi_{em}^*(q) \, \phi_{gn}(q) \, dq \right|^2 \quad (12.19)$$

$$F_{m0} = e^{\frac{-sS^m}{m!}} \tag{12.20}$$

The overlap integral between the vibrational states is

$$S = \frac{W_{LR}}{\hbar\omega_v} \tag{12.21}$$

in which $S = 0$ and 1 mean no overlap and complete overlap, respectively.

When $S \gg 1$ and the vibrational excited state is classically treated, Eq. 12.18 is in the Condon approximation as

$$A(\hbar\omega) \propto e^{\frac{(W_0 - \hbar\omega)^2}{4k_B T' W_{LR}}} \tag{12.22}$$

in which

$$k_B T' = \frac{\hbar\omega_v}{2} \coth\left(\frac{\hbar\omega_v}{2k_B T}\right) \tag{12.23}$$

The linewidth of absorption $A(\hbar\omega)$ is proportional to $\sqrt{\coth\left(\dfrac{\hbar\omega_v}{2k_B T}\right)}$.

Equation 12.22 reproduces quite well the absorption spectra of phonon bands when phonon side-bands dominate the line profile in a very wide temperature range from low to high temperatures. The integrated intensity of the absorption line μ is

$$\int A(\hbar\omega)\, d\hbar\omega = |\mu_{eg}|^2 \tag{12.24}$$

which is independent of the temperature T. This fact shows the validity of the Condon approximation.

It is noted that in the case of the second- and higher-order perturbation by the interaction of electrons with nuclei, we can assume that U_g is not necessarily equal to U_e. Namely,

$$U_g(q) = a\, q^2 \quad (a > 0) \tag{12.25}$$

$$U_e(q) = W_0 + (a - b)q^2 \quad (b > 0) \tag{12.26}$$

At sufficiently high temperature, if $\sigma = \dfrac{a}{b}$, Eq. 12.18 becomes

$$A(\hbar\omega) = A(\hbar\omega_0)\, e^{-\frac{\sigma(\hbar\omega_0 - \hbar\omega)}{k_B T}} \tag{12.27}$$

Equation 12.27 describes the profile of the absorption tail to the lower energy side and is called *Urbach tail* (Fig. 12.7).

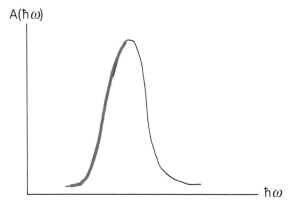

Figure 12.7   Urbach tail.

Here σ, ω_0, $A(\hbar\omega_0)$ are independent of the temperature T. The Urbach tails are associated with the absorption bands of many localized electronic centers in solids and liquids such as ionic crystals, organic crystals, and semiconductor crystals.

12.5   Origin of Spectral Profile

The origins of the energy broadening of the electronic levels can be explained in terms of homogeneous and inhomogeneous broadening. On one hand, in homogeneous broadening, if the damping coefficient γ denotes the sum of the transition probabilities from a certain level to all the other states, the level will have an energy broadening of $\hbar\gamma$. This broadening energy is determined by the uncertainty principle as

$$\Delta E \Delta t \geq \frac{\hbar}{2} \tag{12.28}$$

It is also called lifetime broadening. If the phase factor ϕ jumps at random at the rate γ, the energy spectrum of the level is a Lorentzian with a full width at half maximum (FWHM) of 2γ.

When the gas molecules collide and electronic transition occurs or the molecules decompose or dissociate, there is a broadening of the electronic levels. When there is no electronic transition or no molecular decomposition or dissociation, but there is an effect of phase disturbance, the spectral line also broadens. This effect is enhanced in proportion to the gas pressure. The former effect is called collision broadening, and the latter effect is called pressure broadening.

On the other hand, in inhomogeneous broadening, when there is a distribution of crystal fields due to lattice imperfection or impurities, the energy of localized electronic states has a distribution, and consequently energy broadening of the levels occurs. This is called stress broadening.

A molecule approaching and going away from an observer emits luminescence photons at different wavelengths. This leads to a Doppler broadening of the photon emission of gas molecules.

If a molecule emits photons at different wavelengths depending on the surrounding fields, the linewidth of the emitted photons becomes sharp when the molecule begins to move. This is because all the moving molecules see the same averaged surrounding field. This effect is called motional narrowing.

12.6 Problems

1. Explain the definition of electronic, vibration, and rotation energy level.
2. Discuss the Frank–Condon principle.
3. Prove that the absorption spectrum can be written as

$$A(\hbar\omega) \propto e^{-\frac{(W_0 - \hbar\omega)^2}{4k_B T W_{LR}}}$$

4. Explain the concepts of Born–Oppenheimer approximation and Condon approximation.
5. Show that the width of the absorption spectrum in

$$A(\hbar\omega) \propto e^{\frac{(W_0 - \hbar\omega)^2}{4k_B T' W_{LR}}}$$

for $S \gg 1$ is proportional to $\sqrt{T}$ for high temperature and constant for low temperature.

6. Show that the curvatures of the adiabatic potentials of the electrons of ground state and excited state are not necessarily the same if the interaction between the electrons and the nuclei has terms higher than the square of q.

7. Prove that

$$A(\hbar\omega) = A(\hbar\omega_0)e^{-\frac{\sigma(\hbar\omega_0 - \hbar\omega)}{k_B T}}$$

is valid for the case of the second- and higher-order perturbation by the interaction of electrons with nuclei at sufficiently high temperature.

8. Discuss Urbach's rule.

9. Draw roughly the spectrum of any solid material affected by the vibration of the molecules.

10. Discuss why lifetime energy is broaden based on the uncertainty principle, and also discuss the phenomenon called motional narrowing in NMR and ESR.

Chapter 13

Optical Spectra of Materials

13.1 Introduction

In this chapter, we will discuss and analyze the absorption spectra due to a localized hydrogen-like electronic state or an electronic state forming a band structure in a solid.

13.2 Optical Spectra of Atoms

A hydrogen atom has one proton with one electron, as shown in Fig. 13.1.

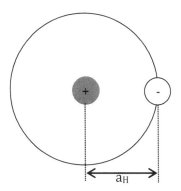

Figure 13.1 Hydrogen atom.

Optical Properties of Solids: An Introductory Textbook
Kitsakorn Locharoenrat
Copyright © 2016 Pan Stanford Publishing Pte. Ltd.
ISBN 978-981-4669-06-1 (Hardcover), 978-981-4669-07-8 (eBook)
www.panstanford.com

The Schröedinger equation is

$$\left(\frac{-\hbar^2 \nabla^2}{2\mu_H} - \frac{e^2}{r}\right)\varphi(\vec{r}) = W\varphi(\vec{r}) \tag{13.1}$$

in which the reduced mass is

$$\mu_H = \frac{mM}{m + M} \tag{13.2}$$

where m and M are electron mass and proton mass, respectively.

A solution to Eq. 13.1 with the wave function ψ and energy W is

$$\psi_{nlm}(r, \theta, \phi) = R_{nl}(r)Y_{lm}(\theta, \phi) \tag{13.3}$$

$$W = \frac{\mu_H e^4}{2\hbar^2 n^2} = -\frac{R_H}{n^2} = \frac{-e^2}{a_H} \cdot \frac{1}{n^2} \tag{13.4}$$

where n is the principle quantum number (such as 1s, 2s 2p); l is the azimuthal quantum number ($0 \le l \le n - 1$), and $l = 0, 1, 2, 3, ...$ denote s, p, d, f, ... states, respectively; m is the magnetic quantum number (such as $m = -1, 0, +1$ if $l = 1$); R_H is the Rydberg constant (which is the energy of the 1s electron of hydrogen and is equal to 13.6 eV); a_H is the Bohr radius (which is the radius of the 1s electron orbital of hydrogen and is equal to 0.529 Å).

Also $\hat{l}$ is the orbital angular momentum operator with eigenvalue $= \sqrt{l(l+1)}\hbar$ or $m_l\hbar$ for $\hat{l}_z$; $\hat{s}$ is the spin angular momentum operator with eigenvalue $= \sqrt{s(s+1)}\hbar$ or $m_s\hbar$ for $\hat{s}_z$; and $\hat{j}$ is the total angular momentum in which $\hat{j} = \hat{l} + \hat{s}$.

The atomic selection rule for the electric dipole transition is

$$\Delta l = \pm 1 \quad \Delta s = 0 \quad \Delta j = 0, \pm 1 \tag{13.5}$$

However, there is also a spin–orbit interaction in the hydrogen atom, as shown in Fig. 13.2.

We will now discuss an atom with more than one electron, as seen in Fig. 13.3.

When the electrons form the closed shells of orbitals, only one configuration of electrons is present. In the case of open shells where orbitals are partly occupied, several eigenstates are associated with one configuration of electrons. These eigenstates take different eigenenergies and are specified by their angular momentum quantum numbers, which we will discuss in two cases.

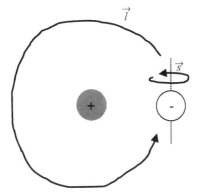

Figure 13.2 Spin–orbit coupling because electron has intrinsic spin s and orbital spin l.

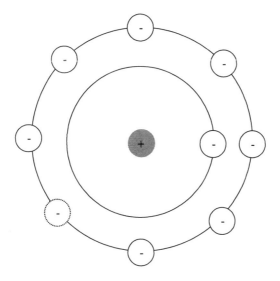

Figure 13.3 Atom with eight electrons.

First, in the case of a light atom with atomic number Z less than 30, the spin–orbit interaction is small and so it is not taken into consideration. Then we specify the electronic states by the total orbital angular momentum of electrons

$$\hat{\vec{L}} = \sum_i \hat{\vec{l}}_i \tag{13.6}$$

and the total spin angular momentum of electrons

$$\hat{\vec{S}} = \sum_i \hat{\vec{s}}_i \tag{13.7}$$

We get the total angular momentum of electrons as

$$\hat{\vec{J}} = \hat{\vec{L}} + \hat{\vec{S}} \tag{13.8}$$

with the total angular momentum quantum number in the integer steps as

$$J = L + S \to |L - S| \tag{13.9}$$

Some nomenclatures are introduced for atom with the new atomic symbol as $^{2s+1}L_J$ in which the terms defined by the quantum numbers L and S are called multiplicity (i.e., singlet, doublet, triplet, etc.) as well as $L = 0, 1, 2, 3, \dots \to s, p, d, f, \dots$

Such a way of specification of electronic configuration is called *LS coupling* or *Russel–Sanderes coupling*. For the energy positions of these fine structures, Hund's rule and Lande's rule hold. As an example, here we study the case of H atom. The 2p orbital with one electron is split into sublevels by spin–orbit coupling, as shown in Fig. 13.4. Two lines are then seen in the 2p → 1s emission spectrum of H atom.

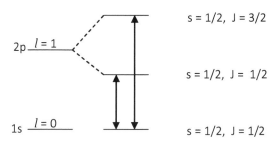

Figure 13.4 Hydrogen atom showing all degenerate states.

Second, in the case of a heavy atom with atomic number Z more than 30, the *LS* coupling scheme becomes worse approximation. The splitting by J value is small for small atomic numbers, so they are called fine structures. This splitting becomes large for large atomic numbers, and they are no longer fine structures. It is more appropriate if we specify the terms by $\overline{\hat{j}_i}$ of i electrons, because the splitting by J becomes larger. Such a way of specification of electronic configuration is called JJ coupling. Then we can specify the electronic states by the total angular momentum of one electron as

$$\hat{\overline{j}}_i = \hat{\overline{l}}_i + \hat{s}_i \tag{13.10}$$

and we get the total angular momentum of i electrons as

$$\hat{\overline{J}} = \sum_i \hat{j}_i \tag{13.11}$$

In addition to the spin–orbit interaction, the crystal field or the electric field induced by neighboring ions also affects the electronic states of the local electron system. The Hamiltonian of this effect is

$$\widehat{H} = \widehat{H}_0 + \widehat{H}_{SO} + \hat{V}_{crystal} \tag{13.12}$$

For example, lanthanide ions have $\widehat{H}_{SO} \gg \hat{V}_{crystal}$, while transition metals have $\widehat{H}_{SO} \ll \hat{V}_{crystal}$.

13.3   Optical Spectra of Molecules

The vibrational and rotational spectra of the molecules are seen under the infrared region 400–4000 cm^{-1} (Fig. 13.5).

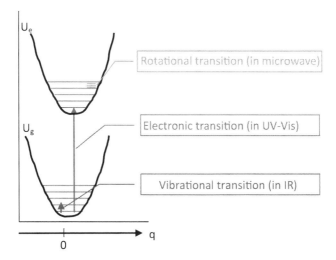

Figure 13.5   Classifications of spectroscopies.

Purely rotational spectra are located in the infrared and microwave regions less than 10 cm^{-1}.

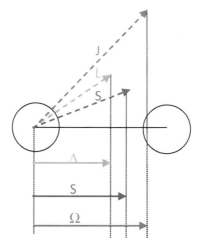

Figure 13.6 Diatomic molecule with momentums along the molecular axis.

In the case of electronic spectra of molecules (Fig. 13.6), new nomenclatures are introduced for linear molecules with new molecular symbol as $^{2S+1}\Lambda_\Omega$ in which Λ is the total orbital angular momentum quantum number along the molecular axis and

$$\Lambda = 0, 1, 2, 3, ... \quad \rightarrow \quad \Sigma, \Pi, \Delta, \Phi, ...$$

Also S is the total spin angular momentum quantum number along the molecular axis, and Ω is the total angular momentum quantum number along the molecular axis. Subscripts g and u are symmetric and antisymmetric, respectively. Superscripts + and − are symmetric and antisymmetric as compared to a mirror plane, including the molecular axis, respectively. The molecular selection rule of the electric dipole transition is

$$\Delta\Lambda = 0, \pm 1 \quad \Delta S = 0 \quad \Delta\Omega = 0, \pm 1 \tag{13.13}$$

For S states: $+ \rightarrow +$ and $- \rightarrow -$

For homonuclear diatomic molecules: $g \rightarrow u$

When two atoms in a molecule have electronic levels of similar energies, these two electronic levels form bonding (π or σ) and antibonding (π^* or σ^*) orbitals and the total energy is minimized. Here are some examples. Methane molecule CH_4 has only σ bonds, while the double bond of $CH_2 = CH_2$ has one σ bond and one π bond. The triple bond of $CH \equiv CH$ has one σ bond and two π bonds. The bonding strength of the σ electron is strong, whereas that of the π electron is weak. The overlap of the bonding π electrons is smaller than that of the bonding σ electrons, so the former has smaller binding energy than the latter. Hence, the π electron system has a contribution in the visible range of the optical spectrum.

13.4 Optical Spectra of Crystals

When the electronic wave functions of atoms overlap, electrons become more mobile. In other words, the lifetime of an electron at one atom becomes small and the band broadens. The probability of electronic transition is then written by using Fermi's golden rule as

$$w_{cv} = \frac{2\pi}{\hbar} \frac{\hbar e^2}{2m^2 \omega \varepsilon_0 \Omega} \left| \left\langle \Psi_c \left| \vec{e} \cdot \hat{\vec{p}} \, e^{i(\vec{k}\cdot\vec{r} - \omega t)} \right| \Psi_v \right\rangle \right|^2 \left| \langle n-1 | \hat{a} | n \rangle \right|^2 \times$$

$$\delta(W_c - W_v - \hbar\omega) \qquad (13.14)$$

In detail, if the bottom of a conduction band (c) and the top of a valence band (v) relevant to the fundamental absorption edge of a crystal (i.e., GaAs, CdS, etc.) are located at the same k point in the Brillouin zone, as shown in Fig. 13.7a, we regard the optical transition occurring at the absorption edge as *direct transition*. We call the band gap W_g as the *direct band gap*. In the direct transition, the optical transition occurs almost vertically.

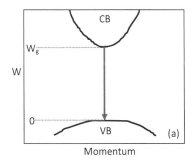
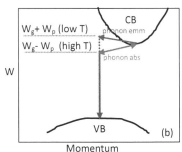

Figure 13.7 *W-k* diagrams showing direct e^-/h^+ recombination (a) and indirect e^-/h^+ recombination (b). CB and VB stand for conduction and valence bands, respectively.

Let the energies of both bands be

$$W_c(\vec{k_c}) = W_g + \frac{\hbar^2 k_c^2}{2m_e} \tag{13.15}$$

$$W_v(\vec{k_v}) = \frac{\hbar^2 k_v^2}{2m_h} \tag{13.16}$$

Then the absorption coefficient (Fig. 13.8) is

$$\alpha(\hbar\omega) = A_1 \sqrt{\hbar\omega - W_g} \tag{13.17}$$

in which

$$A_1 = \sqrt{\frac{\mu^3}{2}} \frac{e^2 \omega |\vec{r}_{cv}|^2}{3\hbar^3 c\pi\varepsilon_0} \quad \text{at } \Gamma \text{ point} \tag{13.18}$$

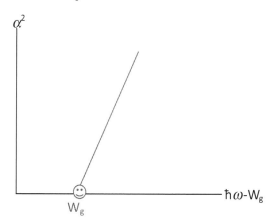

Figure 13.8 Absorption coefficient as a function of energy for direct transition.

If the transition is forbidden at Γ point ($k = 0$), we get the absorption coefficient as

$$\alpha(\hbar\omega) = A_2 (\hbar\omega - W_g)^{3/2} \tag{13.19}$$

On the other hand, as the bottom of a conduction band and the top of a valence band relevant to the absorption process in the problem (i.e., Ge, Si, etc.) are situated at different k points in the Brillouin zone, as shown in Fig. 13.7b, the transition is an *indirect transition*. We call the band gap the *indirect band gap*. The optical absorption is associated with phonon absorption or creation. The absorption coefficient will be

$$\alpha(\hbar\omega) = A_3(\hbar\omega - W_g \mp W_p)^2 \tag{13.20}$$

At $T = 0$, only phonon creation occurs and optical absorption is observed at the energy greater than $W_g + W_p$ (Fig. 13.9). At higher temperature, absorption is associated with phonon emission and optical absorption is observed at the energy greater than $W_g - W_p$.

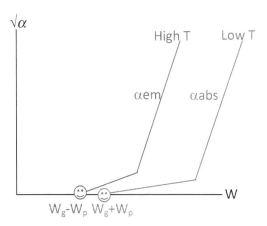

Figure 13.9 Absorption coefficient as a function of energy for indirect transition.

In the discussion of optical properties of solids, one of the most important elementary excitations is *exciton*. This is because it has a large optical transition probability and dominates the optical spectra at low temperatures. In this case, interband transition leaves an electron on the conduction band and a hole on the valence band. The electron and the hole bind each other by Coulomb attraction. They form hydrogen-like energy levels. The exciton absorption band is observed on the lower energy side of the absorption edge. The first exciton level is located at an energy lower than the absorption band by the energy of the exciton binding energy.

Putting κ as the relative dielectric constant, we get the energy and Bohr radius of exciton as

$$W_{ex} = W_g - \frac{R_{ex}}{n^2} + \frac{\hbar^2 k_{ex}^2}{2(m_e + m_h)} \tag{13.21}$$

$$R_{ex} = \frac{\mu_{ex}}{\kappa^2} \cdot \frac{e^4}{2\hbar^2} = \frac{\mu_{ex}}{\mu_H \kappa^2} R_H \tag{13.22}$$

$$a_{ex} = \frac{\mu_H}{\mu_{ex}} \kappa \, a_H \qquad (13.23)$$

Here, μ_{ex} is the reduced mass of the exciton, but it is much smaller than μ_H. The relative dielectric constant κ is large. The binding energy W_{ex} is much smaller than that of hydrogen and the Bohr radius is much smaller. Excitons behave as electrically neutral particles. The exciton absorption series begins with 1s exciton. Note that the 1s absorption of hydrogen atom is not observed.

In Fig. 13.10, the effective radius a of Wannier exciton (i.e., GaAs) is larger than the atomic distances in the crystal (a). This Wannier exciton is observable only at low temperature because it has small binding energies.

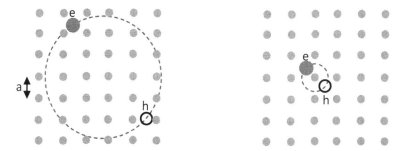

Figure 13.10 Wannier exciton (left) and Frenkel exciton (right).

In Fig. 13.10, the effective radius of Frenkel exciton (alkali halides KBr, NaCl, LiF and rare gas crystals Ne, Ar, Kr, Xe) is smaller than the atomic distances in the crystal (a). This exciton can be regarded as an excitation within an atom. The difference between a Frenkel exciton and a simple electronic excitation within an atom is that the former is mobile.

In addition, when an electron or a hole can be confined under a structure called *quantum well*, as shown in Fig. 13.11, the electron or the hole has the eigenenergy

$$W_n = \frac{\pi^2 \hbar^2 n^2}{2mL^2} \qquad (13.24)$$

An effect of this kind is called the *quantum size effect*. In addition to quantum wells, super-lattices and quantum wires are also much studied topics.

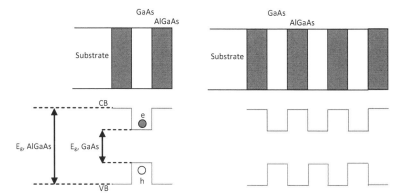

Figure 13.11 Quantum wells: single quntum well (left) and super-lattice (right).

13.5 Problems

1. Prove that in a hydrogen atom, the Schröedinger equation is

$$\left(\frac{-\hbar^2\nabla^2}{2\mu_H} - \frac{e^2}{r}\right)\varphi(\vec{r}) = W\varphi(\vec{r})$$

2. Discuss the relations between principle quantum number, azimuthal quantum number, and magnetic quantum number.

3. Explain the selection rules for multiple atoms taking into account the spin–orbit coupling.

4. Discuss Hund's rule, Lande's rule, and selection rule.

5. Discuss the difference between LS and JJ coupling.

6. Explain the optical spectra of the energy levels of Na, Ca, and Sn.

7. The wave function ψ of the electronic state of a molecule consisting of $N\pi$ electrons is written as

$$\psi = \sum_{j=1}^{N} c_j \phi_j$$

where ϕ is the wave function of each atom. Show that c_j is obtained by solving the equation

$$\sum_{j=1}^{N}(H_{ij} - WS_{ij})c_j = 0$$

Here $H_{ij} = \langle \phi_i | \widehat{H} | \phi_j \rangle$ where $\widehat{H}$ is the Hamiltonian of one π electron as $S_{ij} = < \phi_i | \phi_j >$.

8. Suppose that there are N atoms of the same species aligned in a straight line with constant spacing of a. Suppose also that $\phi(x)$ denotes the wave function of an electron in an isolated atom and

$$\psi(x) = \sum_n c_n \, \phi(x - na)$$

is the wave function of an electron in the lattice. Show that

$$\psi_k(x + ma) = e^{ikma} \, \psi_k(x)$$

must hold and $c_n = Ne^{ikna}$ in order that the wave function ψ_k satisfies the Bolch theorem

$$\psi_{\vec{k}}(\vec{r}) = e^{i\vec{k}\cdot\vec{r}} u_{\vec{k}}(\vec{r})$$

9. Explain the direct and indirect band transition.

10. Discuss the difference between Wannier exciton and Frenkel exciton.

Chapter 14

Some Interesting Phenomena

14.1 Introduction

In this chapter, we will go over some well-known phenomena currently under research. We will also deal with current topics on the optical second-harmonic spectroscopy of solid surfaces. The nonlinear optical response of materials enables us to perform surface-sensitive spectroscopy.

14.2 Luminescence

Luminescence is a process in which an excited electronic state of a molecule makes a transition to the lower energy state emitting fluorescence or phosphorescence, as shown in Fig. 14.1. Fluorescence has a lifetime of several nanoseconds, whereas phosphorescence has a lifetime longer than that of fluorescence (μs-ms). There are many types of luminescence: cathodoluminescence, electroluminescence, and chemiluminescence. Cathodoluminescence is induced by electron beam excitation, while electroluminescence is induced by applying electric field to the material. Chemiluminescence is induced by chemical reaction.

The luminescence intensity with respect to the photon energy of exciting light is called the excitation spectrum. It is basically the

Optical Properties of Solids: An Introductory Textbook
Kitsakorn Locharoenrat
Copyright © 2016 Pan Stanford Publishing Pte. Ltd.
ISBN 978-981-4669-06-1 (Hardcover), 978-981-4669-07-8 (eBook)
www.panstanford.com

same as the absorption of the same material. The luminescence peak in Fig. 14.2 is located on the lower energy (higher wavelength) side of the absorption spectrum by the energy called *Stokes shifts*.

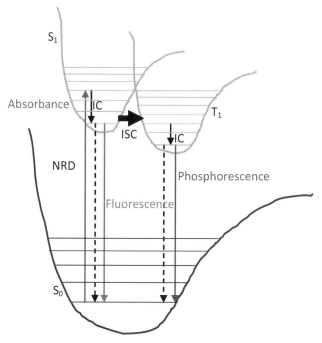

Figure 14.1 Energy level diagram. IC: internal conversion; ISC: intersystem crossing; NRD: nonradiative decay from phonon or lattice vibration.

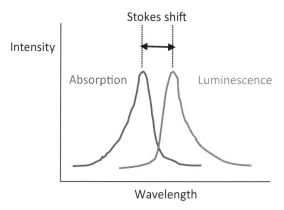

Figure 14.2 Stokes shifts.

The line shapes of luminescence and absorption are mirror images of one another. Namely, the line shape functions of absorption $A(\hbar\omega)$ and luminescence $F(\hbar\omega)$ are related by

$$A(\hbar\omega) = F(2W_e - \hbar\omega) \tag{14.1}$$

Where W_e is the energy of the zero point line.

In Fig. 14.3, when the shapes of the potentials of the excited and ground states of the adiabatic potential are different, the absorption line has a long tail described by Urbach's rule.

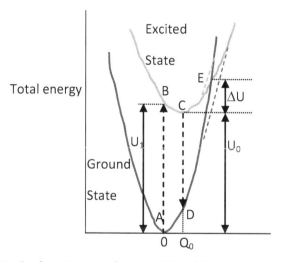

Figure 14.3 Configuration coordinate model and luminescent transition.

At low temperatures, the electron transfer from point B to point C becomes slow and an electronic transition can occur before the electron reaches point C. The energy of the luminescence becomes higher when the temperature becomes lower. This phenomenon is called *dynamic Stokes shifts*. On the other hand, the quantum efficiency of the photon emission according to the scheme shown in Fig. 14.3 decreases as the temperature is raised. At high temperatures, the electron does not follow the path from C to D, but goes to point E and returns to point A. This is called *nonradiative transition*. The quantum efficiency η and the lifetime τ of the luminescence are

$$\eta = \frac{A}{A + s\,e^{-\frac{\Delta W}{k_B T}}} \tag{14.2}$$

$$\tau = \frac{1}{A + s\, e^{-\frac{\Delta W}{k_B T}}}$$ (14.3)

From the temperature dependence of η and τ, one can obtain the transition probability A of the luminescence, frequency factor s, and activation energy ΔW.

Actually the crossing at E in Fig. 14.3 does not occur when the two electronic levels interact. In this case, the electron makes a nonradiative transition by tunneling to the lower electronic state.

In a semiconductor, a crystal of high purity luminescence from a free exciton is observed; however, luminescence is normally observed from a bound exciton. The free exciton is trapped by some impurity centers in a crystal and becomes the bound exciton. This bound exciton disappears radiatively and emits luminescent light. The bound exciton shows stronger luminescence than the free exciton. Thus, the intensity of this bound exciton can be the measure of the purity of the crystal. Moreover, for a neutral donor and $m_h \gg m_e$, the binding energy is one third that of the donor, while if $m_h \ll m_e$, the binding energy is one eighth that of the donor (Fig. 14.4). For a neutral acceptor and $m_h \ll m_e$, the binding energy is one third that of the acceptor, whereas if $m_h \gg m_e$, the binding energy is one eighth that of the acceptor.

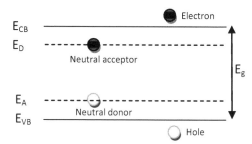

Figure 14.4 Energy level diagram of a donor–acceptor pair and an electron–hole pair.

When the impurity element and one of the elements constituting the bulk material belong to the same column in the periodic table, this impurity becomes a trap of excitons. For example, GaP:N is a semiconductor with an indirect band gap, but the electrons trapped in N centers are localized and have an ambiguity of wave vector *k*. Hence, the direct transition is allowed at the indirect gap and luminescence is observed. In the case of GaP:ZnS, luminescence occurs due to the overlap of the wave functions of the donor and the acceptor.

14.3 Light Scattering

Light can be scattered when it is emitted by a material in all directions as soon as the material absorbs light. Light scatters when there is an inhomogeneity in a medium. The conservation laws can be graphically described in Fig. 14.5 and are written as

$$\omega_s^\pm = \omega_0 \pm \omega \tag{14.4}$$

$$\vec{k}_0 \pm \vec{q} - \vec{k}_s^\pm = 0 \tag{14.5}$$

$$\left| \vec{k}_s^\pm \right| = \frac{n\omega_s^\pm}{c} \tag{14.6}$$

The *Stokes* and *anti-Stokes* components are the "+" and "–" components, respectively. In Eqs. 14.4–14.6, $\omega_s^\pm$ and $\vec{k}_s^\pm$ are the frequency and the wave vector of the scattered light, respectively; ω and $\vec{k}$ are the frequency and the wave vector of the elementary excitation, respectively; ω_0 is the frequency of the incident light; n is the refractive index; and c is the velocity of light.

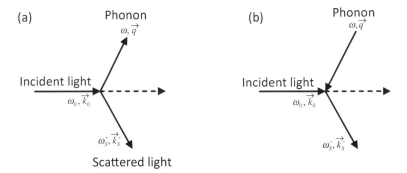

Figure 14.5 Stokes (a) and anti-Stokes scattering (b).

From Eqs. 14.4 to 14.6, we can say that if $\omega = 0$, it is called *Rayleigh scattering*, as seen in Fig. 14.6. When $\omega \approx 0$, it is called *Brillouin scattering* caused by sound wave or acoustic phonon.

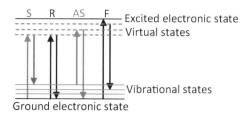

Figure 14.6 Jablonski diagram showing Raman and Rayleigh scattering. S: Stokes; R: Rayleigh; AS: anti-Stokes; F: fluorescence.

For *Raman scattering*, ω is large, and this scattering is caused by molecular vibration, optical phonon, or electronic level in a crystal. The relation between the scattered electric field E_s and the incident electric field E_p is

$$E_s = R_{\rho\sigma} E_\rho \tag{14.7}$$

Using the Kramers–Heisenberg relation for a molecule or local electronic center in a material, one can get the Raman tensor as

$$R_{\rho\sigma} = \sum_m \left[\frac{\langle f|M_\rho|m\rangle\langle m|M_\sigma|g\rangle}{\hbar(\omega_{mg} - \omega_0)} + \frac{\langle f|M_\sigma|m\rangle\langle m|M_\rho|g\rangle}{\hbar(\omega_{mf} + \omega_0)} \right] \tag{14.8}$$

Using Fermi's golden rule, the scattering probability is written as

$$w = \frac{2\pi}{\hbar} \left| \sum_{M_1 M_2} \frac{\langle F|\hat{H}_{\text{int}}|M_2\rangle\langle M_2|\hat{H}_{\text{int}}|M_1\rangle\langle M_1|\hat{H}_{\text{int}}|I\rangle}{(W_I - W_{M2})(W_I - W_{M1})} \right|^2 \times \delta(W_I - W_F) \tag{14.9}$$

in which the initial electronic state is

$$|I = |g; n_0, n_s, n_q \tag{14.10}$$

and the final electronic state is

$$|F = |g; n_0 - 1, n_s + 1, n_q \pm 1 \tag{14.11}$$

Here $\hat{H}$ operates as the electron–photon interaction twice and as the electron–phonon interaction once.

14.4 Laser Action

LASER stands for **L**ight **A**mplification by **S**timulated **E**mission of **R**adiation. Some of the scientists who contributed to the development of laser (Fig. 14.7) are as follows:

- 1916: Einstein—theory of stimulated light emission
- 1954: Townes et al.—MASER
- 1960: Maiman—first solid-state laser (Ruby @ 694 nm)
- 1961: Javan et al.—first gas laser (He–Ne @ 633 nm)

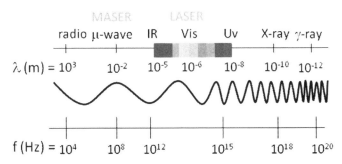

Figure 14.7 Electromagnetic spectrum.

For understanding the interaction of electromagnetic radiation (light) with matter (atom), at least two subjects in physics are needed. One is the *structure of atom* (nucleus consisting of protons and electrons) and its energy states, as shown in Fig. 14.8.

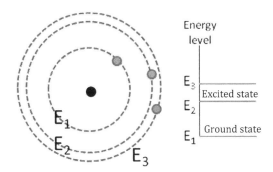

Figure 14.8 Atomic structure and energy level.

Another one is *photon energy* given by

$$E = h(v) = h(c/\lambda) \tag{14.12}$$

$$\Delta E = E_2 - E_1 \text{ (unit: J)} \tag{14.13}$$

where $h = 6.626 \times 10^{-34}$ Js is Planck's constant and 1 eV = 1.6×10^{-19} J/photon, and v is radiation frequency. This formula shows that the energy of each photon is inversely proportional to its wavelength.

This means that each photon of shorter wavelength (i.e., violet light) carries more energy than a photon of longer wavelength (i.e., red light).

Possibly interactions between electromagnetic radiation and matter cause changes in the energy states of the electrons in matter, which can be explained with three mechanisms (Fig. 14.9). First, *stimulated absorption* is an interaction between the incident photon and atom. Second, *spontaneous emission* is the *random* emission of photon by the decay of excited state to a lower level. Last, *stimulated emission* is the *coherence* emission of radiation, stimulated by the photon absorbed by the atom in its excited state. This increase in photon density is called *light amplification*.

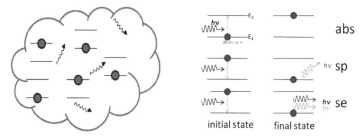

Figure 14.9 Radiative processes. Medium at temperature T (left) and initial/final states (right).

The above-mentioned radiative processes are related to Einstein's coefficients (A, B), as seen in Table 14.1 and Fig. 14.10.

Table 14.1 Radiative processes with Einstein's constants

Radiation	Population rate	Remark(s)
Absorption (abs)	$\left(\dfrac{dN_1}{dt}\right)_{abs} = -B_{12}N_1\rho(v)$	Stimulated emission inversion
Spontaneous emission (sp)	$\left(\dfrac{dN_2}{dt}\right)_{sp} = -A_{21}N_2$	
Stimulated emission (se)	$\left(\dfrac{dN_2}{dt}\right)_{se} = -B_{21}N_2\rho(v)$	$\rho(v)$ = photon density; v = photon frequency

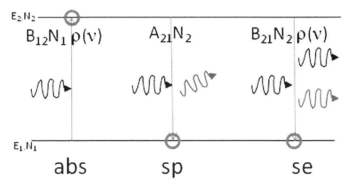

$E_2.N_2$ ────────⊖────────────────────────────

$B_{12}N_1\,\rho(v)$ $A_{21}N_2$ $B_{21}N_2\,\rho(v)$

$E_1.N_1$

abs   sp   se

Figure 14.10 Radiative processes with Einstein's constants.

Suppose a thermal equilibrium at temperature T of matter (atoms between two energy levels) and radiation (photons). That is, the total number of atoms N_2 (N_1) at E_2 (E_1) and photon number will remain constant in time. The total reduction rate of population N_2 is thus

$$\left(\frac{dN_2}{dt}\right) = 0 = -A_{21}N_2 - B_{21}N_2\rho(v) - [-B_{12}N_1\rho(v)] \qquad (14.14)$$

$$\rho(v) = \frac{A_{21}}{B_{12}\left(\dfrac{N_1}{N_2}\right) - B_{21}} \qquad (14.15)$$

Also let the normal population follow Boltzmann distribution at that temperature T (Fig. 14.11) as

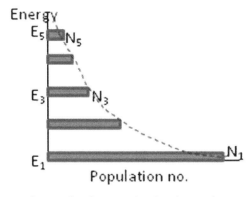

Figure 14.11 Population of each energy level at thermodynamic equilibrium.

$$\rho(\upsilon) = \frac{8\pi h \upsilon^3}{c^3} \cdot \frac{1}{e^{(h\upsilon/kT)} - 1} \tag{14.16}$$

$$\frac{N_2}{N_1} = e^{-\Delta E/kT} = e^{-h\upsilon/kT} \tag{14.17}$$

Substituting Eq. 14.17 in Eq. 14.15, we get

$$\rho(\upsilon) = \frac{A_{21}}{B_{12}e^{(h\upsilon/kT)} - B_{21}} \tag{14.18}$$

Equation 14.18 is equivalent to Eq. 14.16, and we rearrange to isolate the multipliers of the term $e^{h\upsilon/kT}$ to obtain

$$\left(\frac{A_{21}}{B_{21}} - \frac{8\pi h \upsilon^3}{c^3} \cdot \frac{B_{12}}{B_{21}} \right) e^{h\upsilon/kT} = \frac{A_{21}}{B_{21}} - \frac{8\pi h \upsilon^3}{c^3} \tag{14.19}$$

Equation 14.19 must be true if

$$B_{12} = B_{21} \tag{14.20}$$

$$\frac{A_{21}}{B_{21}} = \frac{8\pi h \upsilon^3}{c^3} \tag{14.21}$$

Equation 14.20 indicates the relation between the population rate dN/dt and the total number of atoms N. If $N_1 > N_2$, the rate of stimulated absorption $\left(\dfrac{dN_1}{dt} \right)_{abs} = -B_{12}N_1\rho(\upsilon)$ is high and is called *photon attenuation*. If $N_2 > N_1$, the rate of stimulated emission $\left(\dfrac{dN_2}{dt} \right)_{se} = -B_{21}N_2\rho(\upsilon)$ is high and is called *photon amplification*. Stimulated emission and absorption are, therefore, inverse processes.

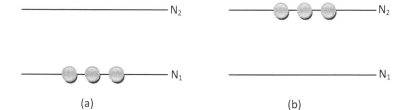

(a)   (b)

Figure 14.12 Normal (left) and inverted population (right).

In the case of $N_2 > N_1$, as shown in Fig. 14.12, this is the condition of *population inversion* for laser action shown in Fig. 14.13—a condition running contrary to the equilibrium population densities predicted by Boltzmann distribution.

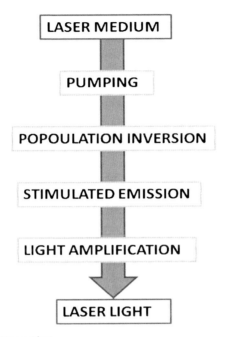

Figure 14.13 Laser action.

Equation 14.21 indicates that

$$\frac{B_{21}}{A_{21}} \propto \frac{1}{v^3} \tag{14.22}$$

which means the lower the frequency (longer wavelength), the higher B_{21}. Here B_{21} relates to stimulated emission, leading to photon amplification. By contrast, A_{21} relates to spontaneous emission, leading to little photon amplification. Hence, laser with short wavelength radiation (i.e., UV or X-ray) is not easy to build and operate (Fig. 14.14).

A laser device consists of three elements, as shown in Fig. 14.15. First is the external energy source or pump. Second is the laser medium or amplifying medium. The last is the optical cavity or resonator.

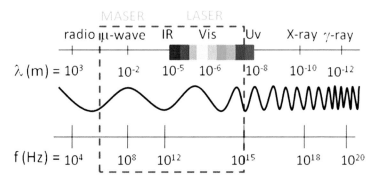

Figure 14.14 Applicable lasers in certain wavelengths.

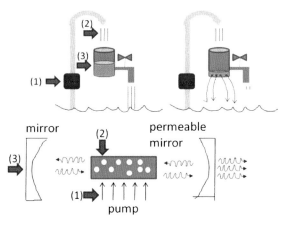

Figure 14.15 Analogy of laser parts.

Pump is the source of energy that excites atoms in an active medium to their excited state to create population inversion. Many pumping methods can be used. One is the electric pumping method used in *gas laser*, for instance in He–Ne laser @ 632.8 nm (Fig. 14.16). A high-voltage power supply causes *electrons* to accelerate from the cathode toward the anode. These electrons collide with He and transfer kinetic energy to the excited He*. After that He* excites Ne.

Another one is the optic pumping method used in *solid* or *liquid laser*, for example in Ruby laser @ 694.3 nm (Fig. 14.17). The *Xe-lamp* is used to excite Cr^{3+} in the ruby rod ($Al_2O_3:Cr^{3+}$).

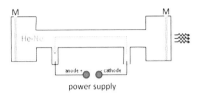

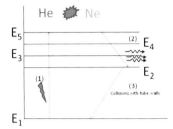

Figure 14.16 He–Ne laser.

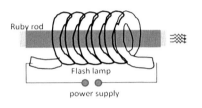

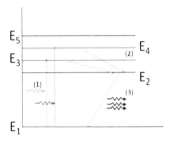

Figure 14.17 Ruby laser.

A *laser medium* is a collection of atoms (or molecules) stimulated to *population inversion*. The medium can be *liquid* (such as Rhodamin 6G Laser) or *gas,* such as He–Ne laser @ 632.8 nm in Fig. 14.16 (visible radiation) and CO_2 laser @ 10.6 μm (IR radiation). In addition, the medium can be solid, such as Ruby laser @ 694 nm and Nd^{3+}:YAG (Yttrium Aluminium Garnet) laser @ 1064 nm (host is YAG crystal rod). Doped material is trivalent Neodymium ions as shown in Fig. 14.18).

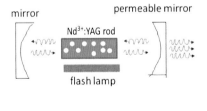

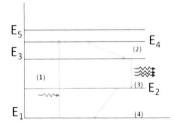

Figure 14.18 Nd^{3+}:YAG laser.

Optical cavity is the space between the laser mirrors in which mirror I has almost 100% reflection, while the output mirror II has partial reflection and transmission (Fig. 14.19). These allow each photon to pass (back and forth) many times through the active medium so that enough amplification of light results. They result in the beam waist W and divergence angle ϕ depending on the mirror distance and the cavity design of radius.

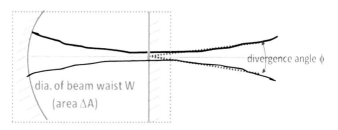

Figure 14.19 The most common optical cavity.

We have so far learned the basic principle of laser. Now how a laser operates? We consider two cases of laser operation: first in terms of energy cycle and second in terms of atom and photon (in cavity).

- **In terms of energy cycle, such as in a four-level laser (Fig. 14.20)**:

  First, pump transition is the creation of photons to excite atoms from E_1 to E_4. Second is rapid nonradiative transition in which atoms from E_4 decay and pile up at E_3. Third is light amplification in which photons reflect axially mirror pairs to stimulate other atoms, thus to emit their photons from E_3 to E_2. Fourth is rapid nonradiative transition in which atoms from E_2 decay to E_1.

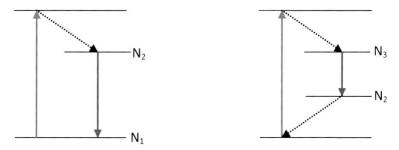

Figure 14.20 Three- (left) and four-level (right) lasers.

- **In terms of atoms and photons in cavity (Fig. 14.21):**
  First is pumping on in which atoms between the mirrors in cavity are raised to the excited state. Second is spontaneous emission in which random emissions of photons leave through the sides of the cavity. In the meantime, stimulated emission also exists. Seed photons along the optical axis of laser reflect off the mirrors to stimulate other atoms to emit their photons. Third is external laser beam in which a fraction of photons with enough intensity pass out through the output mirror.
  The laser behavior can be explained in terms of monochromaticity (wavelength, frequency), coherence (phase), directionality (parallel beam), intensity (brightness), and focusability (a tiny spot).

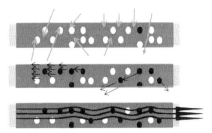

Figure 14.21 Laser operation. Light from the flash tube excites atoms in the ground state (top). Excited atoms emit light waves moving parallel to the mirror axes (middle). Laser light bursts via permeable mirror (bottom).

The laser behavior can be explained in terms of monochromaticity (wavelength, frequency), coherence (phase), directionality (parallel beam), intensity (brightness), and focusability (a tiny spot).

Monochromaticity: Laser radiation is almost one wavelength (one color), as can be measured by the very *narrow linewidth* (Fig. 14.22).

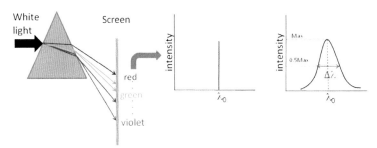

Figure 14.22 Light passing via prism (left). Line shape of laser radiation in theory (middle) and reality (right).

$$E_2 - E_1 = hv \quad \lambda_0 = \frac{c}{v} \tag{14.23}$$

where λ_0 is the central wavelength; $\Delta\lambda$ = FWHM (wavelength) linewidth, measured at 0.5Max of the line shape plot; and Δv = frequency linewidth.

Coherence: Beams are in phase in both time and location (Fig. 14.23).

$$L_c = c\, t_c \quad t_c \approx \frac{1}{\Delta v} \quad \Delta v = \frac{c\,\Delta\lambda}{\lambda_0^2} \tag{14.24}$$

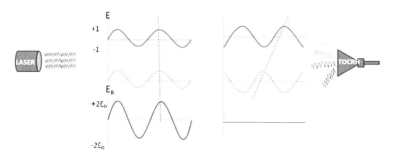

Figure 14.23 Light waves from laser (left) and incandescent lamp (right).

Coherence time t_c is the average time interval over which one can continue to predict the correct phase of laser beam at a given point in space. Coherence length L_c is the average length of the light beam along which the phase of wave remains unchanged.

Directionality: Beam is almost a *parallel beam* and moves in one direction in space as can be measured by the *very small divergence* (Fig. 14.24).

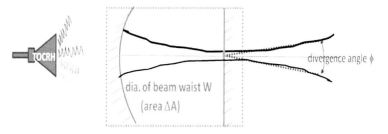

Figure 14.24 Output of incandescent lamp (left) and laser (right).

Divergence angle ϕ is the angle of beam spread given by

$$\phi = \frac{1.27\lambda}{w} \text{ (unit: radian)} \tag{14.25}$$

where w is the beam waist. In Fig. 14.25, the beam waist of Nd^{3+}:YAG laser @ 1064 nm ($w = 3.0$ mm) is bigger than that of He–Ne laser @ 632.8 nm ($w = 0.5$ mm).

Figure 14.25 Beam width with respect to beam angle.

Intensity: Intensity can be explained in terms of output rate and irradiance.

$$\text{Photon output rate} = \frac{1}{\lambda^2} \cdot \frac{1}{e^{(h\upsilon/kT)} - 1} \cdot \Delta A \, \Delta \upsilon \tag{14.26}$$

For example, the photon output rate of 1 mW He–Ne laser is 10^{16} photons per second, while the photon output rate of Hg lamp is 10^9 photons per second.

Irradiance (power density) or radiation power at the illuminated area is expressed as

$$L_e = \frac{4\Phi_e}{\pi\phi^2 \Delta A} \text{ (unit: W/m}^2) \tag{14.27}$$

where Φ_e is laser power and ΔA is the area of beam waist. For example, the power density of Nd^{3+}:YAG laser is $10^9 - 10^{12}$ W/cm^2.

Focusability: Focusability is ability of the beam to be focused down to a small point via positive lens (Fig. 14.26).

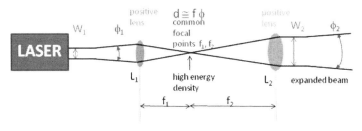

Figure 14.26 Beam expander concept.

Spot size of focused beam can be defined as

$$d \cong f \phi \tag{14.28}$$

where d is the width of the collimated beam, ϕ is the beam divergence angle, and f is the focal length of the lens. However, the beam divergence of laser can be further reduced by using *beam expander* under the condition of $f_2 > f_1$.

Due to

$$w_1\phi_1 = w_2\phi_2 \text{ and } d = f_1\phi_1 = f_2\phi_2 \tag{14.29}$$

the beam divergence of the expanded beam is

$$\phi_2 = \frac{w_1}{w_2}\phi_1 = \frac{f_1}{f_2}\phi_1 \tag{14.30}$$

The beam expansion ratio will be

$$\frac{w_2}{w_1} = \frac{f_2}{f_1} \tag{14.31}$$

Lasers can be classified in many ways, as summarized in Table 14.2 and Fig. 14.27.

Table 14.2   Laser classification

Classification	Examples
Active medium	Gas, solid, liquid lasers
Emission wavelength	IR, Vis, UV
Output power	0.1 mW to 600 W
Beam spot	0.4 mm to 4 cm
Beam divergence	0.2–18 mrad

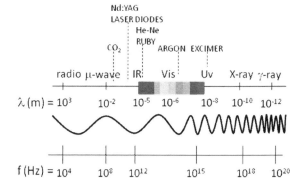

Figure 14.27 Principal wavelengths of common lasers.

Laser applications can be classified into two divisions: (1) laser and interaction and (2) laser and information. They are summarized in Table 14.3.

Table 14.3   Overview of laser uses

Laser and interaction	Laser and information
Industrial applications	Information processing
Medical uses	Remote sensing
Scientific applications	Alignment facilitator
Marking	Holography
Laser spectroscopy	Scanner
Weaponry	Entertainment

An example of lasers and interaction is the use of lasers in industrial applications (Table 14.4). The laser energy is absorbed by the materials, which raises the local temperature so that the matter could be cut and welded (Fig. 14.28).

Table 14.4   Laser for industrial applications

Laser power	Function
High power: CO_2 laser and Nd^{3+}:YAG laser (via *robot unit*)	General cutting and welding
Low power: He–Ne laser	Easily handling and focusing

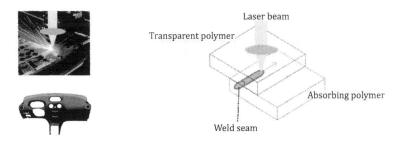

Figure 14.28 Lasers for cutting and welding.

Another example of lasers and interaction is the use of lasers in medical applications. The laser energy absorbed by the biological tissue raises the local temperature, so the tissue can be coagulated or cut. Thermal changes depend on temperature level. Many kinds of changes can occur in biological systems, as shown in Table 14.5. Laser types with specific uses are summarized in Table 14.6 and Fig. 14.29.

Table 14.5 Thermal level versus changes in biosystems

Thermal level	Changes in biosystems
Body temperature till 60°C	Tissue becomes warm
60–65°C	Coagulation
65–90°C	Protein denaturization
90–100°C	Elimination of fluids (Drying)
More than 100°C	Vaporization

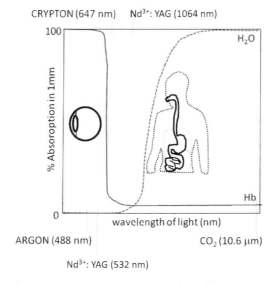

Figure 14.29 Laser types with respect to wavelengths.

In addition, lasers can be used for research applications, such as in nuclear reaction by

Deuterium + Tritium = ^{4}He (3.5 MeV) + n (14.1 MeV)

Table 14.6   Lasers for medical applications

Laser types	Purposes
CO_2 laser highly absorbed by H_2O	General cutting
Nd^{3+}:YAG laser (via fiber-optic scope)	Photocoagulation of ulcer
Excimer laser (i.e., ArF, KrF)	Clean cutting
He–Ne laser	Easy handling and focusing
Ar ion laser highly absorbed by pigmentation	Photocoagulation of retina

The nuclear reaction has four processes and is shown in Fig. 14.30. First, energies from the laser beams (i.e., CO_2 laser, Nd^{3+}:YAG lasers) is circumferentially supplied to a target. Second, the existing waves from the first process compress (D + T) pellets to create a high pressure inside the target. Third, a temperature rises at a center of the target. Last, enormous energy releases from the target on time scale of μs-ps.

laser beam

Figure 14.30 Nuclear reaction.

In information processing, lasers are used in data storage/readout and printing systems. For data storage, laser (such as GaAs laser) creates pits on a photoreactive surface. Data pattern is then kept in binary digit (*bit*) as 1 (land) or 0 (pit). Eight bits is 1 byte or 1 letter or 2 numbers. Bytes 01000001 and 01000010 may represent A and B, respectively. A storage capacity of 4 GB on a 14" CD can translate into 320,000 letter-sized documents. For data readout (Fig.

14.31), laser is focused onto a surface. Light is reflected back from pit and land to the detector into original data pattern.

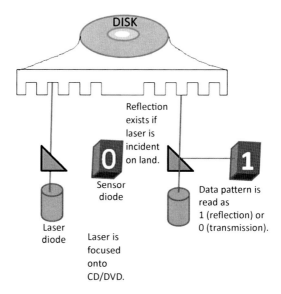

Figure 14.31 Data readout.

Lasers also have a key role in printing systems, as shown in Fig. 14.32. *Traditional printers* (typewriter or dot printer) create mechanical image by pressing ink ribbon (from a formed character of typewriter or from a set of pins of dot printer) onto paper. *Laser printers* create electrostatic images by laser beam scanning across the selective discharge of a photoconductive drum. Toner with charges of opposite polarity adheres to the photoconductive surface and is then transferred onto paper.

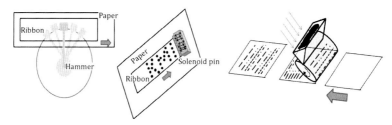

Figure 14.32 Typewriter (left), dot printer (middle), and laser printer (right).

14.5   Optical Second-Harmonic Generation

When an electron is in an ideally parabolic potential, it makes a linear response to the external electric field. When an electron is in a nonparabolic potential, its response to the external field is nonlinear. In a nonlinear optical effect, the magnitude of a generated polarization is not proportional to the magnitude of the incident electric field. We can write the nonlinear polarization compared to the incident electric field as

$$\vec{P} = \varepsilon_0 \left[\chi^{(1)} \vec{E} + \chi^{(2)} \vec{E}^2 + \chi^{(3)} \vec{E}^3 + \cdots \right] \tag{14.32}$$

$$\vec{P} = \vec{P}^{(1)} + \vec{P}^{(2)} + \vec{P}^{(3)} + \cdots \tag{14.33}$$

$$\vec{P} = \vec{P}^{(1)} + \vec{P}^{NL} \tag{14.34}$$

Here $\chi^{(1)}$, $\chi^{(2)}$, $\chi^{(3)}$,... are the nonlinear susceptibility tensors and $\vec{P}^{NL}$ is the nonlinear polarization. For centrosymmetric crystals, we get $\chi^{(2)} = 0$.

For the second-order nonlinear response, the ith component of polarization is

$$P_i^{(2)}(\omega_i) = \sum_{jk} \varepsilon_0 \chi_{ijk}^{(2)} E_j(\omega_j) E_k(\omega_k) \tag{14.35}$$

Due to $\vec{E}(\omega_j) \propto e^{-i\omega_j t}$ $\tag{14.36}$

$$\vec{P}(\omega_i) \propto e^{-i\omega_i t} \tag{14.37}$$

There are many types of frequency mixing processes:

- sum-frequency generation (SFG)

$$\omega_3 = \omega_1 + \omega_2 \tag{14.38}$$

- difference-frequency generation (DFG)

$$\omega_3' = |\omega_1 - \omega_2| \tag{14.39}$$

- second-harmonic generation (SHG)

$$\omega_3 = 2\omega_1 \tag{14.40}$$

Figure 14.33 is an example of the second-harmonic intensity patterns of metallic nanowires analysis using Eq. 14.35.

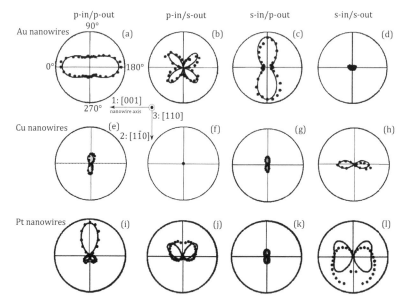

Figure 14.33 Measured (filled circles) and calculated (solid line) second-harmonic intensity patterns for nanowires versus the sample rotation angle ϕ. The fundamental photon energy is 1.17 eV for Au and Cu nanowires and 2.23 eV for Pt nanowires. ϕ is defined as the angle between the incident plane and the [001] direction on the sample surface. The incident angle is 45°.

From the Maxwell's equations,

$$\vec{\nabla} \times \vec{E} = -\frac{\partial \vec{B}}{\partial t} \tag{14.41}$$

$$\vec{\nabla} \times \vec{B} = \mu_0 \frac{\partial}{\partial t}(\tilde{\varepsilon}\vec{E} + \vec{P}^{NL}) \tag{14.42}$$

we get

$$\nabla^2 \vec{E} - \tilde{\varepsilon}\mu_0 \frac{\partial^2 \vec{E}}{\partial t^2} = \frac{1}{\varepsilon_0 c^2}\frac{\partial^2 \vec{P}}{\partial t^2} \tag{14.43}$$

If we consider a second-order nonlinear polarization created by an incident field under the x-direction via a nonlinear susceptibility element $\chi_{yxx}^{(2)}$, the generation of the second-harmonic light wave can be described by the following relations:

$$E_{1x}(z,t) = \tilde{E}_{1x}e^{i(k_1 z - \omega t)} \tag{14.44}$$

$$P_y^{(2)}(z,t) = \varepsilon_0 \chi_{yxx}^{(2)} \tilde{E}_{1x}^2 e^{2i(k_1 z - \omega t)} \tag{14.45}$$

$$E_{2y}(z,t) = \tilde{E}_{2y}(z) e^{i(k_2 z - 2\omega t)} \tag{14.46}$$

Substituting Eqs. 14.44–14.46 in Eq. 14.43, we get

$$\frac{dE_{2y}(z)}{dz} = \frac{2i\omega^2 \chi_{yxx}^{(2)} \tilde{E}_{1x}^2}{c^2 k_2} e^{iz(2k_1 - k_2)} \tag{14.47}$$

Solving Eq. 14.47 with the boundary condition $I_{2y}(z) = 0$ at $z = 0$, we have

$$I_{2y}(z) \propto I_{1x}^2 \frac{\sin^2 \dfrac{\Delta k z}{2}}{\left(\dfrac{\Delta k}{2}\right)^2} \tag{14.48}$$

Here $\Delta k = 2k_1 - k_2$, and it is equal to $2\omega (n_1 - n_2)/c$ where n_1 and n_2 represent the refractive indices of the media at the optical frequencies of ω and 2ω, respectively. $I_{2y}(z)$ is an increasing function of z when z is small. But when z becomes larger than $\pi/\Delta k$, it becomes a decreasing function of z. This is because the phase velocities of nonlinear polarization wave and electromagnetic wave at 2ω are different, and the harmonic waves radiated by various points of polarization wave cancel each other by negative interference. When $\Delta k = 0$, the harmonic waves generated at various points in the material have the same phase, and they show positive interference. In this case, we can say that the phase-matching condition is satisfied.

The optical SHG is a phenomenon in which the polarization

$$P = \chi^{(2)} E^2 \tag{14.49}$$

induced by an incident electric field $E \propto e^{-j\omega t}$ radiates a light of doubled frequency $E \propto e^{-2j\omega t}$. SHG occurs in a medium whose structure lacks inversion symmetry.

Let us consider a potential of an electron of mass m along the coordinate z:

$$\frac{V}{m}(z) = \frac{1}{2}\omega_0^2 z^2 + a_3 z^3 + a_4 z^4 + \cdots \tag{14.50}$$

In Fig. 14.34a, when the potential is symmetric compared to the point $z = 0$, we have $a_{2n+1} = 0$.

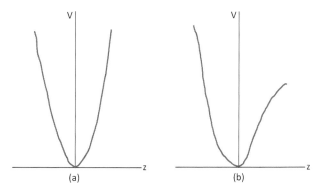

Figure 14.34 One-dimensional potentials with symmetry (a) and without inversion symmetry (b).

If this potential is asymmetric, as shown in Fig. 14.34b, the equation of motion of the electron is

$$m\frac{d^2z}{dt^2} + m\omega_0^2 z + 3a_3 z^2 + 4a_4 z^3 + \cdots = Fe^{-i\omega t} \tag{14.51}$$

The solution to this equation to the first-order by the perturbation theory is

$$z = z^{(0)} + z^{(1)} \tag{14.52}$$

$$z^{(0)} = \frac{F}{m(\omega_0^2 - \omega^2)} e^{-i\omega t} \tag{14.53}$$

$$z^{(1)} = \frac{-3a_3}{m(\omega_0^2 - 4\omega^2)} \cdot \frac{F^2}{m^2(\omega_0^2 - \omega^2)^2} e^{-2i\omega t} \tag{14.54}$$

14.6 Problems

1. Show that there should be six terms in the absolute symbol in the Raman scattering probability as

$$w = \frac{2\pi}{\hbar} \left| \sum_{M_1, M_2} \frac{\langle F|\hat{H}_{int}|M_2\rangle\langle M_2|\hat{H}_{int}|M_1\rangle\langle M_1|\hat{H}_{int}|I\rangle}{(W_I - W_{M2})(W_I - W_{M1})} \right|^2 \times \delta(W_I - W_F)$$

depending on the order in which the absorption and emission of the photons and the phonon, when the initial state and the final state of transition can be given by

$|I\rangle = |g; n_0, n_s, n_q\rangle$

$|F\rangle = |g; n_0 - 1, n_s + 1, n_q \pm 1\rangle$

Which term gives the biggest contribution when $\hbar\omega_0$ is close to the band gap energy?

2. Discuss the difference between normal population and population inversion.

3. Explain the concept of laser oscillation.

4. Prove that

$$\nabla^2 \vec{E} - \tilde{\varepsilon}\mu_0 \frac{\partial^2 \vec{E}}{\partial t^2} = \frac{1}{\varepsilon_0 c^2} \frac{\partial^2 \vec{P}}{\partial t^2}$$

5. Prove that

$$\frac{dE_{2y}(z)}{dz} = \frac{2i\omega^2 \chi_{yxx}^{(2)} \tilde{E}_{1x}^2}{c^2 k_2} e^{iz(2k_1 - k_2)}$$

6. Prove that

$$I_{2y}(z) \propto I_{1x}^2 \frac{\sin^2 \dfrac{\Delta k z}{2}}{\left(\dfrac{\Delta k}{2}\right)^2}$$

7. Show that the resonance wavelength λ is given by

$$m\lambda = 2d\sin\theta$$

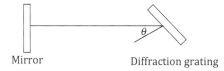

Mirror   Diffraction grating

where d is the period of the grooves on the grating.

8. Show that the peak amplification coefficient of gas with a Lorentzian spectrum is

$$\gamma(\omega_0) = \frac{2\pi c_2 A_{12}}{n^2 \omega_0^2 \Delta\omega} \cdot \frac{\Delta N}{V}$$

where

$$\frac{\Delta N}{V} = \frac{N_2}{V} - \frac{g_2}{g_1} \cdot \frac{N_1}{V}$$

9. Discuss the inversion symmetry of any nonlinear media.

10. Find a solution to

$$m\frac{d^2z}{dt^2} + m\omega_0^2 z + 3a_3 z^2 + 4a_4 z^3 + \cdots = Fe^{-i\omega t}$$

to the first order by the perturbation theory.

Chapter 15

Optics of Eyes

15.1 Introduction

Eye functions like a camera, as shown in Fig. 15.1. Iris allows light into the eye. Cornea, lens, and humors focus light on the retina. The light striking the retina is converted into action potentials relayed to brain.

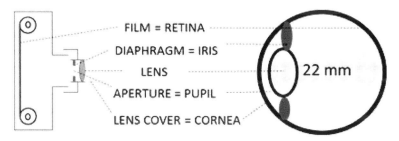

Figure 15.1   Camera–eye analogy.

15.2 Component of Eyes

Figure 15.1 also presents eye components along the optical axis, in the same order hit by light in the *vision process*. Light travels through air (refractive index $n = 1$) and hits the *cornea* ($n = 1.376$) having 12

Optical Properties of Solids: An Introductory Textbook
Kitsakorn Locharoenrat
Copyright © 2016 Pan Stanford Publishing Pte. Ltd.
ISBN 978-981-4669-06-1 (Hardcover), 978-981-4669-07-8 (eBook)
www.panstanford.com

mm ϕ and 0.6 mm thick (near center, and thickening further at the edge). After that light strikes the aqueous humor (n = 1.336). Around this area, there is an *iris* giving an eye pigment and controlling light intensity via an adjustable *pupil*. The light then goes to the lens (n = 1.41), vitreous (n = 1.336), and retina, respectively. At the retina, there are two photoreceptors, as summarized in Table 15.1.

Table 15.1   Details of photoreceptors

Photoreceptor	Quantity	Location	Properties	Visual pigment
Long, thin rod cells	100 millions	Near retina periphery	Insensitive to bright light/color	Purple
Wide cone cells	10 millions	Near retina center	Sensitive to bright light/color	RGB

After retina, the light finally directs to the brain, which fuses two distinct images into one, referred to as *binocular vision*.

15.3   Function of Eyes

There are at least three functions of eyes: accommodation, adaptation, and visual acuity. *Accommodation* is the eye's ability to see objects nearby or far away (Table 15.2 and Fig. 15.2). Accommodation exists for young adults, whereas accommodation loss occurs in the early 40s.

Table 15.2   Accommodation details

Vision	Eyes in	Lens shape	Lens focal length	Lens radius
Near vision	Tensed state	Curve	f = 14 mm	r = 6 mm
Far vision	Relaxed state	Flat	f = 17 mm	r = 10 mm

Adaptation is the eye's sensitivity to light intensity (brightness) regulated at the iris with adjustable pupil (2–8 mm ϕ from bright to dim light).

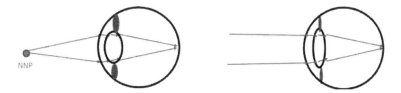

Figure 15.2   Near (left) and far (right) visions. NNP is normal near point.

Visual acuity is the ability to see detailed image (*eye resolution*—the smallest readable letter on an eye chart). Vision test uses a chart of capital letters with various letter sizes on a white background, known as the *Snellen chart*, kept at a fixed testing distance of 20 ft (= 6 m). Visual acuity is measured in terms of the Snellen fraction obtained by dividing the testing distance of defective eyes by the testing distance of normal eyes.

15.4   Ocular Problems and Their Corrections

Ocular problems and their corrections by glasses are summarized in Table 15.3 and Fig. 15.3 and Fig. 15.4.

Figure 15.3   Normal (left), myopia (mid-left), hypermetropia (mid-right), and astigmatism (right).

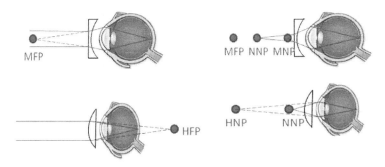

Figure 15.4   Corrections of myopia (left) and hypermetropia (right).

Table 15.3   Ocular problems and their corrections

Ocular problems	Eyeball shape	Symptoms	Corrections (Fig. 5.4)
Normal (N)	Almost sphere	For object from far point (FP), relaxed eyes form a clear vision on retina.	No need
		For object from near point (NP), partially tensed eyes form a clear vision on retina.	
Myopia (M)	Axially too long	For object from ∞FP, relaxed myopic eye forms a blurred vision on retina.	Using negative lens, light from ∞FP appears to originate at MFP.
		For object from MNP, tensed myopic eye forms a clear vision on retina.	Using negative lens, light from NNP appears to originate at MNP.
Hyperopia or hypermetropia (H)	Axially too short	For object from ∞FP, relaxed hyperopic eye forms a clear vision at HFP behind retina.	Using positive lens, light from ∞FP appears to originate at HFP.
		For object from HNP, tensed hyperopic eye forms a clear vision on retina.	Using positive lens, light from NNP appears to originate at HNP.
Astigmatism	Asymmetry of two axes of cornea	Generally blurred vision comes from astigmatism mixed with myopia or hyperopia.	Using sphero-cylindrical lens

When optometrists prescribe corrective glasses for myopic or hyperopic astigmatism (Fig. 15.5), they set three numbers written in prescription format, for instance

R$_x$: −1.0 −2.5 × 90   (myopic astigmatism)

R$_x$: +1.0 −2.5 × 90   (hyperopic astigmatism)

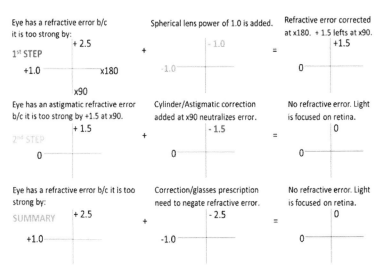

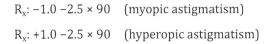

Figure 15.5   Corrections of myopic astimatism.

The ±1 represent the power of a spherical lens (in diopter) to correct myopia or hyperopia. The −2.5 indicates the power of a cylindrical lens (in diopter) to correct astigmatism. Last is the *y*-axis (×90) or *x*-axis (×180) of cylindrical lens.

Not all ocular problems can be corrected with glasses. Some examples are summarized in Table 15.4.

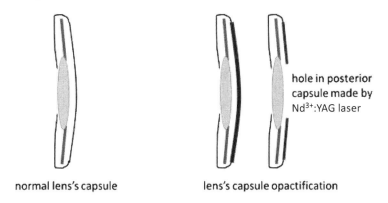

hole in posterior capsule made by Nd^{3+}:YAG laser

normal lens's capsule   lens's capsule opactification

Figure 15.6   Posterior capsulotomy surgery.

Table 15.4 Ocular problems with their therapies

Ocular problems	Symptoms	Therapies
Glaucoma	Increasing fluid pressure in anterior chamber	Nd^{3+}:YAG laser used to open up blocked ducts for better drainage
Diabetic retinopathy	Organic disorders in posterior chamber	Nd^{3+}:YAG laser used to weld the ruptured blood vessel
Radial keratomy	Elongated myopic eyeball needs laser in situ kerotomileusis	MICROKERATOME opens cornea. Next, radial cut by CO_2 laser to get flatten cornea. Last, close cornea.
Posterior capsulotomy (Fig. 15.6)	Regrowth of cells on back wall of lens capsule	Nd^{3+}:YAG laser opens membrane in posterior chamber. Next, remove opacified lens. Last, replace with implant lens.

15.5 Problems

1. Explain the components of the eyes along the optical axis.
2. A person has MFP of 50 cm and MNP of 15 cm. What power of eye glasses is needed to correct MFP from infinity object? What is the new NNP by using this eye glass?
3. A hyperopic person has HNP of 125 cm. What is the power of eye glasses needed to clearly see an object at NNP of 25 cm?
4. The Nd^{3+}:YAG laser used in posterior capsulotomy surgery @ 1.06 µm wavelength has 1 MW power and 0.1 mrad beam divergence at focusing lens (power 20D). (a) What is the spot size of the focused beam on the opaque membrane in the interior of the eye? (b) What is the power density (irradiance) of the focused beam on target?
5. Discuss any ocular problem and its correction.

Appendix A

Waves

A.1 Waves Equation

If we have a stationary wave (*non-sinusoidal* form), as shown in Fig. A.1.1, as

$$y' = f(x') \text{ at } t = 0 \tag{A.1.1}$$

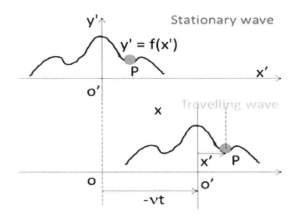

Figure A.1.1 Stationary (top) and traveling waves (bottom).

Optical Properties of Solids: An Introductory Textbook
Kitsakorn Locharoenrat
Copyright © 2016 Pan Stanford Publishing Pte. Ltd.
ISBN 978-981-4669-06-1 (Hardcover), 978-981-4669-07-8 (eBook)
www.panstanford.com

we can have a traveling wave as

$$y = f(x \pm vt) \text{ at } t = t \tag{A.1.2}$$

where $\quad x' = x \pm vt, \quad \dfrac{\partial x'}{\partial x} = 1 \text{ and } \dfrac{\partial x'}{\partial t} = \pm v$

Also when $y = f(x + vt)$, the wave moves in the left direction. When $y = f(x - vt)$, the wave moves in the right direction. From the space derivative as

$$\frac{\partial y}{\partial x} = \frac{\partial f}{\partial x'} \cdot \frac{\partial x'}{\partial x} = \frac{\partial f}{\partial x'} \tag{A.1.3}$$

$$\frac{\partial^2 y}{\partial x^2} = \frac{\partial}{\partial x}\frac{\partial y}{\partial x} = \frac{\partial(\partial y / \partial x)}{\partial x'} \cdot \frac{\partial x'}{\partial x} = \frac{\partial}{\partial x'}\frac{\partial f}{\partial x'} = \frac{\partial^2 f}{\partial x'^2} \tag{A.1.4}$$

and the time derivative as

$$\frac{\partial y}{\partial t} = \frac{\partial f}{\partial x'} \cdot \frac{\partial x'}{\partial t} = \pm v\frac{\partial f}{\partial x'} \tag{A.1.5}$$

$$\frac{\partial^2 y}{\partial t^2} = \frac{\partial}{\partial t}\frac{\partial y}{\partial t} = \frac{\partial(\partial y / \partial t)}{\partial x'} \cdot \frac{\partial x'}{\partial t} = \frac{\partial}{\partial x'}\left(\pm v\frac{\partial f}{\partial x'}\right) \cdot \pm v = v^2 \frac{\partial^2 f}{\partial x'^2} \tag{A.1.6}$$

Equations A.1.4 and A.1.6 yield

$$\frac{\partial^2 y}{\partial x^2} = \frac{1}{v^2} \cdot \frac{\partial^2 y}{\partial t^2} \tag{A.1.7}$$

On the other hand, if we have the wave function of a traveling wave (*sinusoidal waveform*) as

$$y = A \sin k \, (x \pm vt) \tag{A.1.8}$$

the solution to the partial differential equation will be as follows:
Space derivative

$$\frac{\partial y}{\partial x} = kA\cos k(x \pm vt) \tag{A.1.9}$$

$$\frac{\partial^2 y}{\partial x^2} = -k^2 A\sin k(x \pm vt) \tag{A.1.10}$$

Time derivative

$$\frac{\partial y}{\partial t} = \pm kvA\cos k(x \pm vt) \tag{A.1.11}$$

$$\frac{\partial^2 y}{\partial t^2} = -k^2 v^2 A \sin k(x \pm vt) \tag{A.1.12}$$

Equations A.1.10 and A.1.12 also yield

$$\frac{\partial^2 y}{\partial x^2} = \frac{1}{v^2} \cdot \frac{\partial^2 y}{\partial t^2} \tag{A.1.13}$$

On the other hand, a harmonic wave (*sinusoidal form*) can be written as

$$y = A \, {\sin \atop \cos} \, k(x \pm vt) \tag{A.1.14}$$

where A and k are the amplitude and the propagation constant of the periodic wave form/character, respectively. Since the difference between sine and cosine functions is a relative translation of 0.5π radian, it is also enough to treat only one of these wave functions.

For a sine wave of amplitude, as shown in Fig. A.1.2., at constant time t

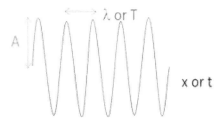

Figure A.1.2 Harmonic waves.

$$y = A \sin k(x + \lambda + vt) = A \sin k(x + vt) + 2\pi \tag{A.1.15}$$

$$A \sin (kx + k\lambda + kvt) = A \sin(kx + kvt + 2\pi) \tag{A.1.16}$$

Clearly,

$$k\lambda = 2\pi \tag{A.1.17}$$

At a constant point x,

$$y = A \sin k(x + v[t + T]) = A \sin k(x + vt) + 2\pi \tag{A.1.18}$$

$$y = A \sin (kx + kvt + kvT) = A \sin (kx + kvt + 2\pi) \tag{A.1.19}$$

Clearly,

$$KvT = 2\pi \tag{A.1.20}$$

Equations A.1.17 and A.1.20 yield

$$v = \frac{\lambda}{T} \text{ and } v = f\lambda \tag{A.1.21}$$

If we define the angular frequency $\omega = \dfrac{2\pi}{T}$, a more compact form of the harmonic wave in Eq. A.1.14 is

$$y = A \, {}^{\sin}_{\cos} \, 2\pi\left(\frac{x}{\lambda} \pm \frac{t}{T}\right) \tag{A.1.22}$$

$$y = A \, {}^{\sin}_{\cos} \, (kx \pm \omega t) \tag{A.1.23}$$

Furthermore, at a constant point x, the phase angle φ is constant or

$$\varphi = k(x \pm vt) \tag{A.1.24}$$

$$d\varphi = 0 = k(dx \pm vdt) \tag{A.1.25}$$

$$\frac{dx}{dt} = \mp v \tag{A.1.26}$$

From Eq. A.1.14, the general form of the harmonic wave at initial phase φ_0, as shown in Fig. A.1.3, can be modified as

$$y = A \, {}^{\sin}_{\cos} \, [k(x \pm vt) + \phi_0] \tag{A.1.27}$$

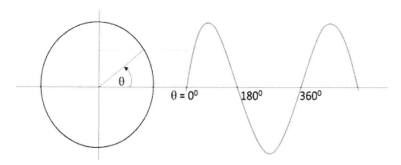

Figure A.1.3 Harmonic waves showing the initial phase.

If we define the initial conditions for harmonic wave at $x = 0$, $t = 0$, and $y = y_0$, we get

$$y_0 = A \, {}^{\sin}_{\cos} \, \varphi_0 \text{ or } \varphi_0 = \sin^{-1}\left(\frac{y_0}{A}\right) \text{ using sine function} \tag{A.1.28}$$

The initial phase φ_0 is generally set equal to zero for simplicity. Let a complex number be defined as

$$\tilde{z} = a + ib \tag{A.1.29}$$

where $a = \text{Re}(z)$, $b = \text{Im}(z)$, and $i^2 = -1$. $\tilde{z}$ can be shown in the polar plot as seen in Fig. A.1.4.

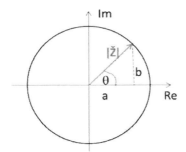

Figure A.1.4 Polar pot of harmonic waves.

The amplitude of $\tilde{z}$ is

$$|\tilde{z}| = \sqrt{a^2 + b^2} \tag{A.1.30}$$

$$a = |\tilde{z}|\cos\theta \text{ and } a = |\tilde{z}|\sin\theta \tag{A.1.31}$$

Substituting Eq. A.1.31 in Eq. A.1.29, we get

$$\tilde{z} = |\tilde{z}|\cos\theta + i|\tilde{z}|\sin\theta = |\tilde{z}|(\cos\theta + i\sin\theta) \tag{A.1.32}$$

By Euler's formula, the expression in the parentheses of Eq. A.1.32 is

$$e^{i\theta} = \cos\theta + i\sin\theta \tag{A.1.33}$$

Equation A.1.32 is modified as

$$\tilde{z} = |\tilde{z}|e^{i\theta} \tag{A.1.34}$$

where $\tan\theta = \dfrac{b}{a}$

The complex conjugate of $\tilde{z}$ is

$$\tilde{z}^* = a - ib = |\tilde{z}|e^{-i\theta} \tag{A.1.35}$$

and

$$\tilde{z}\tilde{z}^* = (|\tilde{z}|e^{i\theta})(|\tilde{z}|e^{-i\theta}) = |\tilde{z}|^2 \tag{A.1.36}$$

Using Euler's formula, the complex form of the harmonic wave is finally expressed as

$$\tilde{z} = |\tilde{z}|\cos\theta + i|\tilde{z}|\sin\theta = |\tilde{z}|e^{i\theta} \tag{A.1.37}$$

$$\tilde{y} = A\cos(kx \pm \omega t) + iA\sin(kx \pm \omega t) = Ae^{i(kx \pm \omega t)} \tag{A.1.38}$$

These show that a calculation using complex form gives correct results for both sine and cosine waves. The complex form of a three-dimensional harmonic wave is expressed as

$$\psi = Ae^{i(kr \pm \omega t)} \tag{A.1.39}$$

Let $k_x x + k_y y + k_z z = kr$ where (k_x, k_y, k_z) are components of the propagation direction and (x, y, z) are components of a point in space. The partial differential equation satisfied by such three-dimensional waves is a generalization of

$$\frac{\partial^2 y}{\partial x^2} = \frac{1}{v^2} \cdot \frac{\partial^2 y}{\partial t^2} \quad \text{(in one dimension)} \tag{A.1.40}$$

in the form

$$\frac{\partial^2 \psi}{\partial x^2} + \frac{\partial^2 \psi}{\partial y^2} + \frac{\partial^2 \psi}{\partial z^2} = \frac{1}{v^2} \cdot \frac{\partial^2 \psi}{\partial t^2} \tag{A.1.41}$$

$$\left(\frac{\partial^2}{\partial x^2} + \frac{\partial^2}{\partial y^2} + \frac{\partial^2}{\partial z^2} \right)\psi = \frac{1}{v^2} \cdot \frac{\partial^2 \psi}{\partial t^2} \tag{A.1.42}$$

If a Laplacian operator is

$$\Delta = \vec{\nabla} \cdot \vec{\nabla} = \vec{\nabla}^2 \equiv \frac{\partial^2}{\partial x^2} + \frac{\partial^2}{\partial y^2} + \frac{\partial^2}{\partial z^2}$$

we get

$$\vec{\nabla}^2 \psi = \frac{1}{v^2} \cdot \frac{\partial^2 \psi}{\partial t^2} \tag{A.1.43}$$

In contrast to a plane wave having a constant amplitude A (i.e., waves on a rope or spring radiating from any positions in the *x-y* plane), a spherical wave (i.e., water or sound wave) propagating further from a point source decreases in amplitude, as shown in Fig. A.1.5.

Figure A.1.5 Waves on rope/spring (left) and water waves (right).

The complex form of a three-dimensional harmonic spherical wave is expressed as

$$\psi = \frac{A}{r} e^{i(kr \pm \omega t)} \tag{A.1.44}$$

If $\theta = kr \pm \omega t$, a plane electromagnetic wave (i.e., light) traveling in an arbitrary direction is explained by the harmonic wave equation as

$$\vec{E} = E_0 e^{i(kr \pm \omega t)} \tag{A.1.45}$$

$$\vec{B} = B_0 e^{i(kr \pm \omega t)} \tag{A.1.46}$$

where $\vec{E}$ and $\vec{B}$ are electric and magnetic field, respectively, shown in Fig. A.1.6. $\vec{E}$ and $\vec{B}$ travel with a constant propagation vector k and frequency ω, and thus with a constant wavelength λ and velocity c.

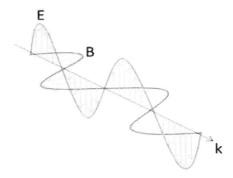

Figure A.1.6 Traveling of the light.

E_0 is electric field amplitude in vacuum and B_0 is magnetic field amplitude in vacuum, in which

$$E_0 = cB_0 \tag{A.1.47}$$

At any specific time and place in any medium,

$$\vec{E} = c\vec{B} \tag{A.1.48}$$

where

$$c^2 = \frac{1}{\varepsilon_0 \mu_0} \tag{A.1.49}$$

ε_0 and μ_0 are the permittivity and the permeability of vacuum, respectively, given by

$\varepsilon_0 = 8.8542 \times 10^{-12}$ $(C.s)^2/(kg.m^3)$ or $C^2/(N.m^2)$ or $C/(V.m)$

$\mu_0 = 4\pi \times 10^{-7}$ $kg.m/(A.s)^2$ or N/A^2

They yield $c = 2.998 \times 10^8$ m/s as the speed of light in vacuum.

While any waves are traveling, they also provide a total energy density u or transmission energy per unit volume (unit: J/m^3) including

$$u_E = \frac{1}{2}\varepsilon_0 \vec{E}^2 \tag{A.1.50}$$

$$u_B = \frac{1}{2\mu_0}\vec{B}^2 \tag{A.1.51}$$

Due to

$$u_B = \frac{1}{2\mu_0}\left(\frac{\vec{E}}{c}\right)^2 = \frac{1}{2\mu_0}(\varepsilon_0\mu_0)\vec{E}^2 = u_E \tag{A.1.52}$$

the total energy (u) of the electromagnetic wave is

$$u = u_E + u_B = 2u_E = 2u_B \tag{A.1.53}$$

$$u = \varepsilon_0 \vec{E}^2 = \frac{\vec{B}^2}{\mu_0} \tag{A.1.54}$$

$$u = \varepsilon_0 c \vec{E}\vec{B} \tag{A.1.55}$$

If we define 'Power' as an energy u per unit time changes Δt (as shown in Fig. A.1.7) or

$$\text{Power} = \frac{u\,\Delta V}{\Delta t} = \frac{uAc\Delta t}{\Delta t} = ucA \quad \text{(unit: J/s-m}^2) \tag{A.1.56}$$

and also define the Poynting vector $\vec{S}$ as the power per unit area or

$$\vec{S} = uc \tag{A.1.57}$$

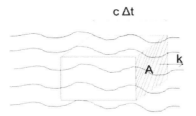

Figure A.1.7 Transmission energy of light via a medium.

Substituting Eq. A.1.57 in Eq. A.1.55, we get

$$\vec{S} = \varepsilon_0 c^2 \vec{E}\vec{B} \tag{A.1.58}$$

Also defining irradiance L_e (unit: W/m^2) as the average value of the Poynting vector $\langle \vec{S} \rangle$ where $\langle \cos^2 \theta \rangle = \dfrac{1}{2}$ in a period, we get

$$L_e = \frac{1}{2}\varepsilon_0 c^2 E_0 B_0 = \frac{1}{2}\varepsilon_0 c E_0^2 = \frac{1}{2\mu_0} c B_0^2 \text{ (in free space)} \tag{A.1.59}$$

It should be noted that the velocity is c/n if the waves are traveling in a medium of refractive index n.

A.2 Superposition of Waves

We learned about single harmonic wave in Section A.1. In Section A.2, we combine two or more harmonic waves existing together along the same direction to form a net displacement called *superposition*. For one harmonic wave, we have

$$\vec{E} = E_0 \sin(kr + \omega t + \varphi_0) \tag{A.2.1}$$

When kr is constant, a constant phase angle α is

$$\alpha = kr + \varphi_0 \tag{A.2.2}$$

Substituting Eq. A.2.2 in Eq. A.2.1, we get

$$\vec{E} = E_0 \sin(\omega t + \alpha) \tag{A.2.3}$$

Applying Eq. A.2.3 for two harmonic waves of identical frequency, we get

$$\vec{E}_1 = E_{01} \sin(\omega t + \alpha_1) \tag{A.2.4}$$

$$\vec{E}_2 = E_{02}\sin(\omega t + \alpha_2) \tag{A.2.5}$$

Using the following trigonometric equations

$$\sin(A + B) = \sin A \cos B + \cos A \sin B \tag{A.2.6}$$

$$\cos(A - B) = \cos A \cos B + \sin A \sin B \tag{A.2.7}$$

Eqs. A.2.4 and A.2.5 are rewritten as

$$\vec{E}_1 = E_{01}\sin(\omega t + \alpha_1) = E_{01}(\sin\omega t \cos\alpha_1 + \cos\omega t \sin\alpha_1) \tag{A.2.8}$$

$$\vec{E}_2 = E_{02}\sin(\omega t + \alpha_2) = E_{02}(\sin\omega t \cos\alpha_2 + \cos\omega t \sin\alpha_2) \tag{A.2.9}$$

The net wave is

$$\vec{E}_R = \vec{E}_1 + \vec{E}_2 = E_{01}\sin(\omega t + \alpha_1) + E_{02}\sin(\omega t + \alpha_2) \tag{A.2.10}$$

$$\vec{E}_R = (E_{01}\cos\alpha_1 + E_{02}\cos\alpha_2)\sin\omega t + (E_{01}\sin\alpha_1 + E_{02}\sin\alpha_2)\cos\omega t \tag{A.2.11}$$

Based on the phasor diagram of two waves (phasor is a vector having size equal to the *amplitude* of wave and rotating around the origin with *phase angle* α), as shown in Fig. A.2.1, we get

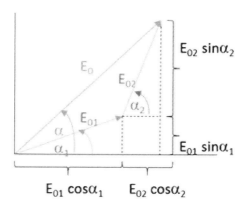

Figure A.2.1 Harmonic waves of the same frequencies.

$$E_0\cos\alpha = E_{01}\cos\alpha_1 + E_{02}\cos\alpha_2 \quad \text{(in x-axis)} \tag{A.2.12}$$

$$E_0\sin\alpha = E_{01}\sin\alpha_1 + E_{02}\sin\alpha_2 \quad \text{(in y-axis)} \tag{A.2.13}$$

Substituting Eqs. A.2.12 and A.2.13 in Eq. A.2.11, we get the resultant waves as

$$\vec{E}_R = E_0 \sin \omega t \cos \alpha + E_0 \cos \omega t \sin \alpha \qquad \text{(A.2.14)}$$

$$\vec{E}_R = E_0 \sin(\omega t + \alpha) \qquad \text{(A.2.15)}$$

with a phase via

$$\tan \theta = \frac{E_{01} \sin \alpha_1 + E_{02} \sin \alpha_2}{E_{01} \cos \alpha_1 + E_{02} \cos \alpha_2} \qquad \text{(A.2.16)}$$

together with an amplitude via

$$E_0^2 = (E_{01} \cos \alpha_1 + E_{02} \cos \alpha_2)^2 + (E_{01} \sin \alpha_1 + E_{02} \sin \alpha_2)^2 \qquad \text{(A.2.17)}$$

$$E_0^2 = E_{01}^2 \cos^2 \alpha_1 + E_{02}^2 \cos^2 \alpha_2 + 2E_{01} \cos \alpha_1 E_{02} \cos \alpha_2 +$$

$$E_{01}^2 \sin^2 \alpha_1 + E_{02}^2 \sin^2 \alpha_2 + 2E_{01} \sin \alpha_1 E_{02} \sin \alpha_2 \qquad \text{(A.2.18)}$$

$$E_0^2 = E_{01}^2 + E_{02}^2 + 2E_{01}E_{02} \cos(\alpha_1 - \alpha_2) \qquad \text{(A.2.19)}$$

Next, we extend from two harmonic waves to i ($i = 1 \to n$) harmonic waves of identical frequencies, and we get the resultant waves with a phase via

$$\tan \alpha = \frac{\sum\limits_{i=1}^{n} E_{0i} \sin \alpha_i}{\sum\limits_{i=1}^{n} E_{0i} \cos \alpha_i} \qquad \text{(A.2.20)}$$

together with an amplitude via

$$E_0^2 = \left(\sum_{i=1}^{n} (E_{0i} \cos \alpha_i \right)^2 + \left(\sum_{i=1}^{n} (E_{0i} \sin \alpha_i)^2 \right) \qquad \text{(A.2.21)}$$

$$E_0^2 = E_{0i}^2 \cos^2 \alpha_i + 2 \sum_{i=1}^{n} \sum_{j>i} E_{0i}E_{0j} \cos \alpha_i \cos \alpha_j +$$

$$E_{0i}^2 \sin^2 \alpha_i + 2 \sum_{i=1}^{n} \sum_{j>i} E_{0i}E_{0j} \sin \alpha_i \sin \beta_j \qquad \text{(A.2.22)}$$

$$E_0^2 = \sum_{i=1}^{n} E_{0i}^2 + 2 \sum_{i=1}^{n} \sum_{j>i} E_{0i}E_{0j} \cos(\alpha_i - \alpha_j) \qquad \text{(A.2.23)}$$

Superposition conditions of equal amplitude (i.e., E_{01}) and frequency from N sources giving N harmonic waves are dependent on the sources.

If N sources are random or out of difference phase (i.e., a flashlight used as a light source) so that $\cos(\alpha_i - \alpha_j)$ becomes 0 as N increases or

$$E_0^2 = \sum_{i=1}^{n} E_{0i}^2 + 2\sum_{i=1}^{n}\sum_{j>i} E_{0i}E_{0j}\cos(\alpha_i - \alpha_j) \qquad \text{(A.2.24)}$$

$$E_0^2 = \sum_{i=1}^{n} E_{0i}^2 = N\,E_{01}^2 \qquad \text{(A.2.25)}$$

$$E_0 = \sqrt{N}\,E_{01} \qquad \text{(A.2.26)}$$

Since the irradiance (W/m^2) is dependent on E_0^2, we get the net irradiance equal to N times (irradiance of individual source).

If N sources are coherent or of difference phase (i.e., laser used as a light source) so that $\cos(\alpha_i - \alpha_j)$ reaches 1 or

$$E_0^2 = \sum_{i=1}^{n} E_{0i}^2 + 2\sum_{i=1}^{n}\sum_{j>i} E_{0i}E_{0j}\cos(\alpha_i - \alpha_j) \qquad \text{(A.2.27)}$$

$$E_0^2 = \sum_{i=1}^{n} E_{0i}^2 + 2\sum_{i=1}^{n}\sum_{j>i} E_{0i}E_{0j} \qquad \text{(A.2.28)}$$

$$E_0^2 \cong \left(\sum_{i=1}^{n} E_{0i}\right)^2 = N^2 E_{01}^2 \qquad \text{(A.2.29)}$$

$$E_0 = N\,E_{01} \qquad \text{(A.2.30)}$$

As the irradiance (W/m^2) is dependent on E_0^2, the net irradiance is equal to N^2 times (irradiance of individual source).

In the case of the back-and-forth waves along the same transmitting medium (two sine waves traveling in opposite directions) with an identical amplitude and frequency, we get standing waves as shown in Fig. A.2.2.

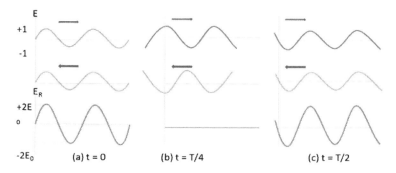

Figure A.2.2 Standing waves at $t = 0$, $0.25T$, and $0.5T$.

$$\vec{E}_1 = E_0 \sin(kx + \omega t) \tag{A.2.31}$$

$$\vec{E}_2 = E_0 \sin(kx - \omega t) \tag{A.2.32}$$

We can get the resultant waves as

$$\vec{E}_R = \vec{E}_1 + \vec{E}_2 = E_0\left[\sin(kx + \omega t) + \sin(kx - \omega t)\right] \tag{A.2.33}$$

Using the following trigonometric equation

$$\sin A + \sin B = 2\sin\left(\frac{A+B}{2}\right) + \cos\left(\frac{A-B}{2}\right) \tag{A.2.34}$$

in Eq. A.2.33, we obtain

$$\vec{E}_R = 2E_0 \sin\left(\frac{kx + \omega t + kx - \omega t}{2}\right) + \cos\left(\frac{kx + \omega t - kx + \omega t}{2}\right) \tag{A.2.35}$$

$$\vec{E}_R = 2E_0 \sin\left(\frac{2kx}{2}\right) + \cos\left(\frac{2\omega t}{2}\right) \tag{A.2.36}$$

$$\vec{E}_R = 2E_0 \sin kx + \cos \omega t \tag{A.2.37}$$

From Eq. A.2.37, the maximum E_R exists when $\cos \omega t = \pm 1$ or

$$\cos(\omega t) = \cos(m\pi) \pm 1 \quad m = 0, \pm1, \pm2, \ldots \tag{A.2.38}$$

$$\frac{2\pi t}{T} = m\pi \tag{A.2.39}$$

We obtain an antinode of standing wave occurring at

$$t = \frac{1}{2}mT = 0, 0.5T, T, 1.5T, \ldots \tag{A.2.40}$$

and we also get a node of standing wave occurring at

$$t = 0.25T, 0.75T, 1.25T, \ldots \tag{A.2.41}$$

in which all amplitudes pass through zero at the fixed nodal point at each 0.5λ.

Superposition of waves of different frequency (Fig. A.2.3) showing wavelength and velocity with identical amplitude is

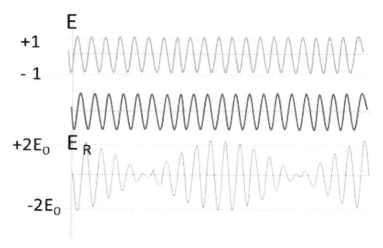

Figure A.2.3 Harmonic waves of different frequencies.

$$\vec{E}_1 = E_0 \cos(k_1 x - \omega_1 t) \tag{A.2.42}$$

$$\vec{E}_2 = E_0 \cos(k_2 x - \omega_2 t) \tag{A.2.43}$$

and we can get E_R by

$$\vec{E}_R = E_0 \left[\cos(k_1 x - \omega_1 t) + \cos(k_2 x - \omega_2 t) \right] \tag{A.2.44}$$

Using the following trigonometric equation

$$\cos A + \cos B = 2\cos\left(\frac{A+B}{2}\right) + \cos\left(\frac{A-B}{2}\right) \tag{A.2.45}$$

and applying Eq. A.2.45 in Eq. A.2.44, we get

$$\vec{E}_R = 2E_0 \left[\cos\left(\frac{k_1 x - \omega_1 t + k_2 x - \omega_2 t}{2}\right) + \cos\left(\frac{k_1 x - \omega_1 t - k_2 x + \omega_2 t}{2}\right) \right] \tag{A.2.46}$$

$$\vec{E}_R = 2E_0 \left[\cos\left(\frac{(k_1 + k_2)x - (\omega_1 + \omega_2)t}{2}\right) + \cos\left(\frac{(k_1 - k_2)x - (\omega_1 - \omega_2)t}{2}\right) \right] \tag{A.2.47}$$

Let

$$\omega_p = \frac{\omega_1 + \omega_2}{2}; \ k_p = \frac{k_1 + k_2}{2} \text{ (higher-frequency wave)} \tag{A.2.48}$$

and

$$\omega_g = \frac{\omega_1 - \omega_2}{2} \; ; \; k_g = \frac{k_1 - k_2}{2} \text{ (lower-frequency wave)} \qquad \text{(A.2.49)}$$

and applying Eqs. A.2.48 and A.2.49 in Eq. A.2.47, we get the *beat phenomena* by the relation

$$\vec{E}_R = 2E_0 \Big[\cos(k_p x - \omega_p t) \cos(k_g x - \omega_g t) \Big] \qquad \text{(A.2.50)}$$

Beat frequency ω_b can be defined as a different frequency for two waves by

$$\omega_b = \omega_1 - \omega_2 = 2\omega_g \qquad \text{(A.2.51)}$$

We know that the general relation of velocity is

$$v = f\lambda = \frac{\omega}{k} \qquad \text{(A.2.52)}$$

We then define the *phase velocity* (velocity of the higher-frequency wave) as

$$v_p = \frac{\omega_p}{k_p} = \frac{\omega_1 + \omega_2}{k_1 + k_2} \cong \frac{\omega}{k} \qquad \text{(A.2.53)}$$

and specify the *group velocity* (velocity of the lower-frequency wave) as

$$v_g = \frac{\omega_g}{k_g} = \frac{\omega_1 - \omega_2}{k_1 - k_2} \cong \frac{d\omega}{dk} \qquad \text{(A.2.54)}$$

If light propagates in a *non-dispersive medium* (i.e., vacuum, air), we can say that

$$v_p = c \qquad \text{(A.2.55)}$$

while

$$v_g = \frac{d\omega}{dk} \qquad \text{(A.2.56)}$$

$$v_g = \frac{d(kv_p)}{dk} \qquad \text{(A.2.57)}$$

$$v_g = v_p + k\frac{dv_p}{dk} \cong v_p \qquad \text{(A.2.58)}$$

By contrast, when light travels in a *dispersive medium* (i.e., water, glass), we can say that

$$v_p = \frac{c}{n} \qquad \text{(A.2.59)}$$

$$dv_p = -\frac{c}{n^2}dn \tag{A.2.60}$$

$$\frac{dv_p}{dk} = -\frac{c}{n^2} \cdot \frac{dn}{dk} \tag{A.2.61}$$

whereas

$$v_g = v_p + k\frac{dv_p}{dk} \tag{A.2.62}$$

Substituting Eq. A.2.61 in Eq. A.2.62, we get

$$v_g = v_p - k\left(\frac{c}{n^2} \cdot \frac{dn}{dk}\right) \tag{A.2.63}$$

Substituting c from Eq. A.2.59 in Eq. A.2.63, we get

$$v_g = v_p - k\left(\frac{nv_p}{n^2} \cdot \frac{dn}{dk}\right) \tag{A.2.64}$$

$$v_g = v_p\left[1 - k\left(\frac{1}{n} \cdot \frac{dn}{dk}\right)\right] \tag{A.2.65}$$

We have learned that

$$k = \frac{2\pi}{\lambda} \tag{A.2.66}$$

$$dk = -\frac{2\pi}{\lambda^2}d\lambda \tag{A.2.67}$$

Substituting Eq. A.2.67 in Eq. A.2.65, we finally get

$$v_g = v_p\left[1 - \left(\frac{k}{n} \cdot \frac{dn}{-\dfrac{2\pi}{\lambda^2}d\lambda}\right)\right] \tag{A.2.68}$$

$$v_g = v_p\left[1 + \left(\frac{\lambda}{n} \cdot \frac{dn}{d\lambda}\right)\right] \tag{A.2.69}$$

Chapter 16

Solid Surface

16.1   Introduction

In this chapter, we will learn the concept of "surface" in the solid phase. In general, surface means the counterpart and the contrasted concept of the word "bulk." That is, three-dimensional periodic treatment is available in the bulk. The surface is the interface between the vacuum (even in a gas phase) and the solid in which its property will be remarkably different from those in the interior of the matter. The interface is a part of the contact between two pieces of matter.

16.2   Atomic Arrangement at Surface

In an ideal surface, such as a clean surface, the atomic arrangement on the plane at the surface is the same as that on the plane in the bulk. By contrast, the arrangement in the clean surface is different from that in a contaminated or dirty surface due to the relaxation and reconstruction processes.

Optical Properties of Solids: An Introductory Textbook
Kitsakorn Locharoenrat
Copyright © 2016 Pan Stanford Publishing Pte. Ltd.
ISBN 978-981-4669-06-1 (Hardcover), 978-981-4669-07-8 (eBook)
www.panstanford.com

16.2.1 Relaxation

In relaxation, there is no change in the two-dimensional atomic arrangement on the surface; however, it is vertically changed, as shown in Fig. 16.1.

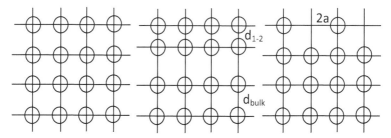

Figure 16.1 Ideal (left), relaxed (middle), and reconstruction (right) surfaces.

16.2.2 Reconstruction

There is a change in the two-dimensional atomic arrangement on the surface, as shown in Fig. 16.1.

16.3 Bravais Lattice

In 1848, Auguste Bravais introduced the five possible lattices in the two-dimensional crystal system, as shown in Fig. 16.2, in which the Bravais lattice is the infinite array of discrete points (i.e., atom, molecule, ion) arranged or oriented in such a way that the array appears the same from whichever point the array is looked at.

16.4 Surface Representation

Generally the surface representation of any solid material A can be simply written as (i.e., Si(111) 7 × 7)

$$A(hkl)\frac{|a_s|}{|a|} \times \frac{|b_s|}{|b|} \tag{16.1}$$

where h, k, l are the plane indices, a is a primitive vector on the ideal surface and a_s is a primitive vector on the reconstruction surface.

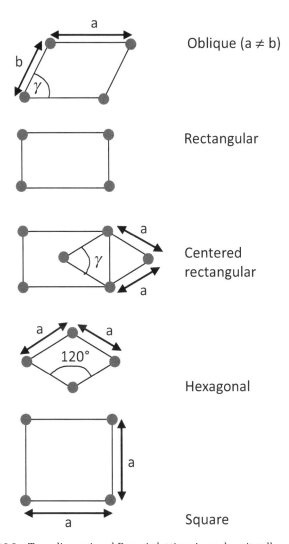

Figure 16.2 Two-dimensional Bravais lattices in each unit cell.

For the ideal surface, the position vector representing a set of basis is

$$R = na + mb \qquad (16.2)$$

where n and m are integers.

After getting the surface reconstruction, the position vector can be modified as

$$R = n(2a) + m(2b) = na_s + mb_s \qquad (16.3)$$

where $2a = a_s$ and $2b = b_s$.

16.5 Work Function

Work function is the energy required to relocate an electron from deep inside the bulk to a point at a distance outside the surface of the bulk. This can be measured from the photoelectric effect, as shown in Fig. 16.3. The work function is usually in the eV scale.

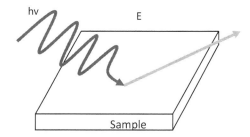

Figure 16.3 Photoelectric effect on the sample surface.

The work function φ relates to the photon energy E as

$$E = hv - \phi \qquad (16.4)$$

where h is Planck's constant and v is the velocity of light.

One can then plot the relation between the work function and the photon energy, as shown in Fig. 16.4.

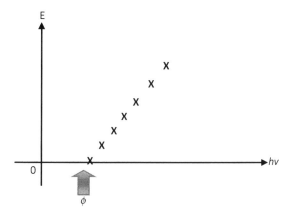

Figure 16.4 Photon energy versus frequency.

The work functions for single crystals at different planes are given in Table 16.1. In general, the highest value of the work function appears on the closed packed surface.

Table 16.1 Measured work functions for single crystals with specified crystal planes

Metal	Crystal plane		
	(110)	**(100)**	**(111)**
Al	4.06	4.41	4.24
Ni	5.04	5.22	5.35
Cu	4.48	4.59	4.98
Ag	4.52	4.64	4.74
Au	5.37	5.47	5.31
Nb	4.87	4.02	4.36
Mo	4.95	4.53	4.55
W	5.25	4.63	4.47

16.6   Problems

1. Discuss the rumpling effect in an ionic crystal.
2. Explain why the number of dangling bonds changes from 49 to 19 after the surface reconstruction of Si(111) 7 × 7.
3. Discuss the differences between the Bravais lattices in two and three dimensions.
4. What is a primitive cell?
5. Why the crystal surface is determined by the surface density?
6. Explain the relation between the Jellium model and the work function.
7. Explain the electron distribution at the surface based on the Jellium model.
8. Discuss a more realistic model based on the muffin-tin potential.
9. Discuss the model of photoelectron emission from the bulk.
10. Explain the potential energy near the surface.

Chapter 17

Scanning Tunneling Microscopy

17.1 Introduction

In scanning tunneling microscopy (STM), we use a scanning tip to monitor the electron transfer at the probe–specimen surface of metals or semiconductors. In general, the probe–specimen gap is about a few nanometers. Fermi levels shift once one supplies the bias voltage to the specimen. Electron then transfers from high to low potential, producing an electrical signal. While scanning, the probe is moved up and down over the specimen to reach a constant electrical signal and sample–tip separation. Tungsten is commonly used as a conductive tip because a sharp tip can be easily produced with tungsten by electrochemical etching techniques.

17.2 Principle

Consider two electrodes separated by a thin barrier from position 0 to S (i.e., vacuum), as seen in Fig. 17.1 (left). Creating a voltage drop between the two metals allows the tunneling current to flow from one metal to another [Fig. 17.1 (right)] as

$$I \propto \int_{0}^{eV} \rho_s(E)\rho_t(E - eV)\,T(E, eV)\,dE \qquad (17.1)$$

Optical Properties of Solids: An Introductory Textbook
Kitsakorn Locharoenrat
Copyright © 2016 Pan Stanford Publishing Pte. Ltd.
ISBN 978-981-4669-06-1 (Hardcover), 978-981-4669-07-8 (eBook)
www.panstanford.com

where ρ_s is the density of state (DOS) of the electron in the specimen at energy E and ρ_t is the DOS of the electron in the probe at energy $E - eV$.

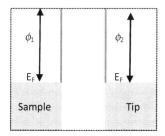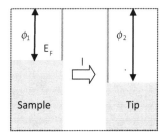

Figure 17.1 One-dimensional potential energy diagram for two parallel metal plates before tunneling (left) and after tunneling (right). E_F represents Fermi energy and φ is the work function (eV).

The transmission probability of finding an electron after the barrier gap S (mostly in Å) is written as

$$T(E,eV) = \exp\left(-2\int_0^S \sqrt{\frac{2m}{\hbar^2}(\phi(z)-E)}\, dz \right) \tag{17.2}$$

where m is the mass of an electron.

The simple potential model of Eq. 17.2 is

$$T(E,eV) = \exp\left(-2S\sqrt{\frac{2m}{\hbar^2}\left(\phi_{avg} - E + \frac{eV}{2} \right)} \right) \tag{17.3}$$

in which the average work function is defined as

$$\phi_{avg} = \frac{\phi_1 + \phi_2}{2} \tag{17.4}$$

If we substitute Eq. 17.3 in Eq. 17.1 and assume that ρ_s and ρ_t are constants, we finally get

$$I = f(V)\exp\left(-1.02 S\sqrt{\left(\phi_{avg} - \frac{|V|}{2} \right)} \right) \tag{17.5}$$

where V is the applied bias voltage. $f(V)$ is the Fermi function containing the weighted joint local density of states and this material property can be obtained by measurements.

From Eq. 17.5, if S is 1 Å and ϕ_{avg} is 5 eV, the current I will change by a factor of 10. Therefore, this shows that when the tip closes enough to the sample surface, we can simply create the tunneling current, although there is a break in the circuit between the barrier gap.

17.3   Implementation

Figure 17.2 shows the block diagram of an STM system. It consists of two main components: a tip and a feedback system.

17.3.1   Scanner

The sharp *tip* is attached onto a piezoceramic element (i.e., tripod, tube, or shear in shape). It is then placed very close to the *sample*. The motion of the probe–specimen in the *xy*-axis is mostly manipulated by a piezoelectric crystal tube.

17.3.2   Feedback System

After scanning the probe over the sample, the electrical signal is transferred to a *feedback circuit*. This system is used to control the gap of the probe–specimen in the *z*-direction.

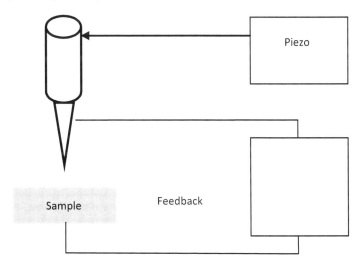

Figure 17.2   Block diagram of an STM system.

If we can keep a constant electrical signal, we will get a constant gap of probe–specimen. In addition, if we can increase the tunneling current when we are on the top of an atom by lowering the tip a little, the attractive force between the tip and the atom would then increase, allowing us to "drag" atoms around. For example, an atom of iron can stick on the copper surface. The other iron atoms can then be dragged along the copper surface to form a circle.

17.4   Strong and Weak Points

STM has some strong and weak points, as summarized in Table 17.1.

Table 17.1   Strong and weak points of STM

Strong points	Weak points
No damage to the sample	Samples limited to conductors and semiconductors
Vertical resolution superior to SEM	Limited to biological applications
Relatively low cost	Generally a difficult technique to perform

17.5   Problems

1. At low voltages, the electrical signal is

   $$I \propto \exp(-2Kd)$$

   $$K = \frac{\sqrt{2m\Phi}}{\hbar}$$

   where d is the tip–sample distance, K is the decay rate, m is the electron mass, Φ is the barrier height, and $\hbar$ is Planck's constant.

   Assume the local barrier height to be 4 eV. Discuss the electrical sensitivity on the gap of the probe–specimen if the current is kept within 2%.

2. Write the one-dimensional potential diagram for STM.

3. Give some examples of STM feedback circuits.

4. Explain the typical piezoscanner.

5. Why do we need coarse positioning and the vibration isolation stage for STM operation?
6. Explain the main components of STM.
7. Discuss the relation between the electron transfer and the gap of probe–specimen.
8. Discuss the atom manipulation in STM.
9. How can we measure the separation and deformation of STM?

Chapter 18

Atomic Force Microscopy

18.1   Introduction

In this chapter, we will learn measurement of nonconducting surfaces using atomic force microscopy (AFM) instead of STM. This technique uses the repulsive and attractive forces between the tip (mounted on a soft cantilever) and the sample. The movement of the cantilever (frequency shift and damping) can be detected by the input laser and the output photodiode systems.

18.2   Principle

In noncontact AFM, as shown in Fig. 18.1, the cantilever is always vibrated by a self-driven circuit at its resonance frequency. The equation of the cantilever motion in noncontact AFM can be written as

$$m\frac{d^2z}{dt^2} + \frac{m\omega_0}{Q}\cdot\frac{dz}{dt} + m\omega_0^2 z = F_{ext} + F_{int} \tag{18.1}$$

where m is the cantilever mass, z is the cantilever displacement, ω_0 is the resonance frequency of the free cantilever, ω is the resonance frequency of the system, Q is the cantilever constant, F_{ext} is the

Optical Properties of Solids: An Introductory Textbook
Kitsakorn Locharoenrat
Copyright © 2016 Pan Stanford Publishing Pte. Ltd.
ISBN 978-981-4669-06-1 (Hardcover), 978-981-4669-07-8 (eBook)
www.panstanford.com

driving force to oscillate the cantilever, and F_{int} is the tip–sample force.

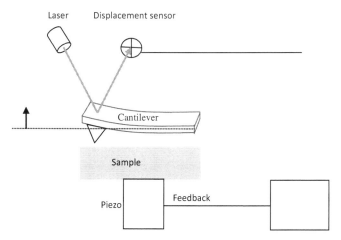

Laser Displacement sensor

Cantilever

Sample

Piezo Feedback

Figure 18.1 Configuration of AFM.

If we assume that the oscillation of the cantilever is harmonic, the solution to Eq. 18.1 is

$$z = A \cos(\omega t) + \cdots \tag{18.2}$$

in which the constant amplitude A is

$$\sqrt{\langle z^2 \rangle} \approx \frac{A}{\sqrt{2}} \tag{18.3}$$

The exciting force to oscillate the cantilever can be written as

$$F_{ext} = F_0 \cos(\omega + a) + \cdots \tag{18.4}$$

If we define

$$\alpha = \frac{\pi}{2} \pm 2n\pi \tag{18.5}$$

Eq. 18.4 becomes

$$F_{ext} = F_0 \cos\left(\omega t + \frac{\pi}{2} \pm 2n\pi\right) + \cdots \tag{18.6}$$

$$F_{ext} \approx -F_0 \sin(\omega t) + \cdots \tag{18.7}$$

The interaction force between the tip and the sample can be written as

$$F_{int}(s) = -\frac{dU(s)}{ds} \tag{18.8}$$

where the interaction potential (van der Waals force) is defined as

$$U(s) \approx \frac{1}{s^{12}} - \frac{AR}{6s} \tag{18.9}$$

where R is the tip radius and s is the separation between the tip and the sample. The frequency shift is finally written as

$$\Delta f \approx -\frac{1}{(2\pi)^2 mA} \int_0^{2\pi/\omega} F_{int}(s)\cos(\omega t)\,dt - \frac{F_0}{4\pi m\omega_0 A}\cdot\cos\alpha \tag{18.10}$$

$$\Delta f \propto -\frac{1}{4\pi m\omega_0} \cdot \frac{dF_{int}(s)}{ds} \tag{18.11}$$

The damping $F_0 A\pi \cdot \sin(\omega t)$ is finally defined as the external force F_{ext} used to keep the oscillation amplitude constant or

$$F_0 A\pi \cdot \sin(\omega t) \approx \frac{\frac{1}{2}m\omega_0^2 A^2 \cdot 2\pi}{Q} + \omega A \int_0^{2\pi/\omega} F_{int}(s)\cdot\sin(\omega t)\,dt \tag{18.12}$$

Potential, force, and frequency shift are plotted in Fig. 18.2.

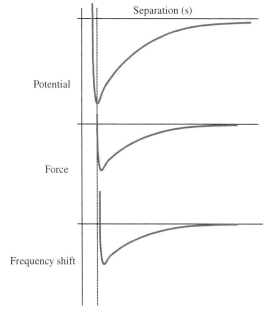

Figure 18.2 Potential (U at the top), force (F at the middle), and frequency shift (Δf at the bottom) at the cantilever oscillation.

18.3   Implementation

Figure 18.1 displays the operation of AFM, which consists of two main parts: scanner and the feedback loop.

18.3.1   Scanner

The piezomodulator is used to fix the probe to control the gap of the probe–specimen (z-axis). On the other hand, the specimen is fixed on a piezoelectric crystal tube (xy-plane). After that the cantilever moves during specimen scanning. The laser light hits around the top of the cantilever end where the probe is set. Next, the displacement sensor detects any reflected light.

18.3.2   Feedback System

After scanning the probe over the sample, the motion of the piezoelectric crystal tube is transferred to a *feedback circuit*. This system is used to control a constant force at the gap of the probe–specimen in the z-direction.

18.4   Strong Points and Weak Points

Some strong and weak points of AFM are summarized in Table 18.1.

Table 18.1   Strong and weak points of AFM

Strong points	Weak points
Samples can be nonconductive	Not easy for alignment
Can be used in air, vacuum, or liquids	

18.5   Problems

1. Which forces relate in AFM?
2. Explain the concept of the piezomodulator.
3. Write the equation of the cantilever in contact AFM.
4. Show the schematic of AFM operation.

5. Explain the damping term.
6. Demonstrate the noncontact AFM control system in a block diagram.
7. Discuss the van der Waals force integrated over the whole volume of the tip and sample.
8. Give some applications of AFM in medical uses.
9. Calculate the amplitude of a 17 KHz force acting on the cantilever for an applied AC voltage of 10 V rms (frequency 17 KHz). The cantilever dimensions are 30 and 100 microns, respectively. It is held 2 μm above the sample. The work function of the cantilever is 5.65 eV, and that of the sample is 5.15 eV.

Chapter 19

Electron Microscopy in Scanning Mode

19.1 Introduction

With scanning electron microscopy (SEM), we can monitor the specimen surface through an electron beam at high power. The electron–specimen interaction produces an SEM image showing atomic structure and orientation.

19.2 Principle

We first spot the probe beam onto the sample (Fig. 19.1) using an accelerated voltage from the electron source. We can use an electron gun via thermionic emission or field electron emission processes to produce the probe beam. Field electron emission relies on the tunneling effect when we use a strong electric field to induce electrons at the surface of the tip. Thermionic emission is based on the heat effect when we rise the temperature of the filament till the emission of electrons at the surface of the filament. Then a primary electron beam dislocates the secondary electron deep below the surface and the backscattered electron in a deeper region. The electron signals are then converted to SEM images via a CCD camera.

Optical Properties of Solids: An Introductory Textbook
Kitsakorn Locharoenrat
Copyright © 2016 Pan Stanford Publishing Pte. Ltd.
ISBN 978-981-4669-06-1 (Hardcover), 978-981-4669-07-8 (eBook)
www.panstanford.com

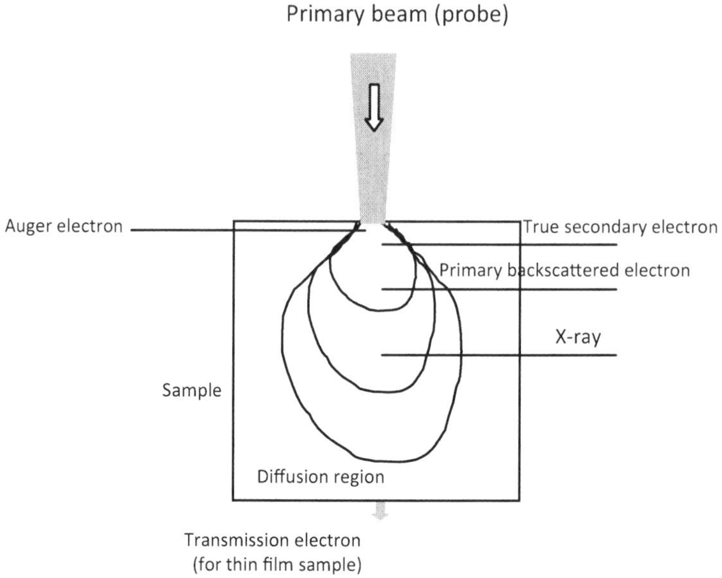

Primary beam (probe)

Figure 19.1 Electron beam hits onto the sample surface.

The obtained current density J can be calculated from Richardson's formula as

$$J = RT^2 e^{-\Phi/kT} \quad \left[\frac{A}{m^2} \right]$$

(19.1)

where T is the temperature of the filament, Φ is the work function, and k is the Boltzmann constant. Richardson's constant is

$$R = \frac{4\pi m q k^2}{h^3} = 1.20 \times 10^6 \quad \left[\frac{A}{(mT)^2} \right]$$

where m is the electron mass, q is the electric charge, and h is Planck's constant.

The current density from the probe indicates the amount of dislodged electrons. That is, the secondary electron (released mostly from specimens of low atomic density, i.e., biological specimens) appears up to a definite limit if the probe energy is high. Above this limit, the secondary electron disappears although the probe energy gets bigger.

The probe also produces the backscattered electron (released mostly from specimens with high atomic density, i.e., metals). The energy of the backscattered electron is higher than that of the secondary electron. An emission above 50 eV is categorized as the reflected (or backscattered) electron. This electron is directly proportional to the atomic number of elements. If the difference of the atomic numbers of two elements is more than 3, one element can be distinguished from the other. For instance, copper (atomic number = 29) shows good contrast on aluminum (atomic number = 13) substrate. In addition, since the backscattered electron reflects from a certain path of travel, another detector is required in addition to the secondary electron detector.

The accelerated voltage V relates to the penetration depth d_p as

$$d_p = \frac{D W V^2}{Z \rho} \tag{19.2}$$

where D is the dimensionless constant, W is the atomic weight, Z is the atomic density, and ρ is the electron density.

The image resolution power r is related to the wavelength of the beam λ as

$$r = \frac{0.612 \lambda}{n \sin \theta} \tag{19.3}$$

where n is the medium refractive index and θ is (1/2) radiation angle from the front of the objective lens.

If the numerical aperture (NA) of the objective lens is defined as

$$NA = n \sin \theta \tag{19.4}$$

then we finally get the resolution power as

$$r = \frac{0.612 \lambda}{NA} \tag{19.5}$$

We can determine the wavelength of the beam based on Broglie's concept. For a particle of mass m with velocity v, the particle wavelength λ is

$$\lambda = \frac{h}{mv} \tag{19.6}$$

where h is Planck's constant (= 6.628×10^{-27} erg-s).

For electron rest mass $m_0 = 9.1 \times 10^{-31}$ kg and velocity v accelerated by low voltage V ($eV = 0.5m_0 v^2$), the electron wavelength λ is

$$\lambda = \frac{h}{m_0 v} = \frac{h}{\sqrt{2m_0 eV}} = \frac{12.3}{\sqrt{V}} \ [\text{Å}] \tag{19.7}$$

If we define $n = 1$ and $\sin\theta \approx \theta$ and we substitute Eq. 19.7 in Eq. 19.5, we finally get

$$r = \frac{7.5}{\sqrt{V}\,\theta} \ [\text{Å}] \tag{19.8}$$

19.3 Implementation

The configuration of SEM consists of four main parts: gun, lens system, image formation, and recorder (Fig. 19.2).

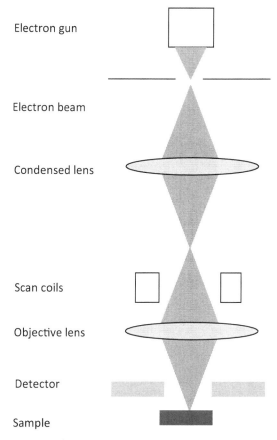

Figure 19.2 SEM configuration.

We first vent the chamber using the nitrogen line. Then we introduce a well-prepared specimen into the chosen holder within the chamber. The specimen (up to 15 cm height) must be electrically conductive and small enough to fit in the vacuum chamber. After that we gently close the chamber door and wait for a vacuum of approximately 7×10^{-5} mBar. Next we can turn the beam source on (5–10 kV). Finally, we can look for the image. At the end, we turn the beam off and vent to remove the specimen. Then we close the chamber door and re-pump.

19.3.1 Gun

A well-prepared specimen is first put in a vacuum chamber. The generated electron with high energy appears after we apply a high voltage from the electron gun. Then the anode position allows the electrons to interact with the specimen. There are two types of electron guns: thermionic and field emission (Table 19.1). A field emission gun is generally better than a thermionic one in terms of brightness, monochromatic electron source, and resolution. In thermionic emission, we can use tungsten or lanthanum hexaboride (LaB_6) served as a filament. In terms of performance, as the work function of LaB_6 is lower than that of tungsten, LaB_6 has more ability to emit electrons than tungsten when equal energy is used. If we require a high-resolution power, LaB_6 is preferred. This is because since the tip of LaB_6 is smaller than that of tungsten, the electron beam crossover has a smaller diameter than tungsten.

Table 19.1 Three types of electron source operating at 100 kV

Properties	LaB_6	Tungsten	Field emission
Work function (eV)	2.4	4.5	4.5
Temperature (K)	1700	2700	300
Current density (A/m^2)	10^6	5×10^4	10^{10}
Crossover size (µm)	10	50	< 0.01
Brightness (A/m^2 sr)	5×10^{10}	10^9	10^{13}
Vacuum (Pa)	10^{-4}	10^{-2}	10^{-8}

19.3.2 Lens System

The electrons are compressed into a beam after they pass through the condensed lens. The scanning coils move the focused beam across the sample in the raster scan pattern. The scan speed is controllable. After that the objective lens focuses the beam spot onto the specimen.

19.3.3 Image Formation

As the electron beam strikes the sample, secondary electrons are emitted from the sample. In addition, backscattered electrons are also emitted from the sample.

19.3.4 Image Recorder

All the electron signals are finally detected and imaged to the CRT camera. Figure 19.3 shows an example of SEM images from tungsten nanotips.

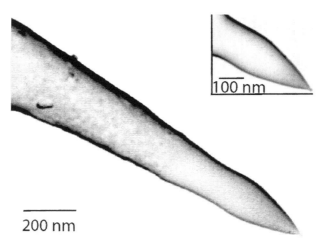

Figure 19.3   SEM images of tungsten nanotips.

19.4   Strong Points and Weak Points

Some strong and weak points of SEM are summarized in Table 19.2.

Table 19.2   Strong and weak points of SEM

Strong points	Weak points
Depth of focus is good.	Many detectors are required.
Shows three-dimensional images.	Sample dimension is limited.

19.5   Problems

1. Discuss the difference between conventional microscopy and SEM.
2. How to prepare a sample for SEM?
3. Explain the mechanism of electron when the electron beam strikes the sample surface.
4. Discuss the advantages and disadvantages of SEM and TEM.
5. Why is SEM with a field emission gun preferred to traditional SEM?
6. How do you avoid charging issues in an insulated sample in SEM?
7. What is the suitable method to obtain the lattice spacing and chemical composition of a 10 nm diameter InN nanowire?
8. Calculate the de Broglie wavelengths of electrons in SEM (30 keV) and TEM (300 keV).

Chapter 20

Electron Microscopy in Transmission Mode

20.1 Introduction

With transmission electron microscopy (TEM), we can monitor the composition and structure of a very small object, at least in the microscale level, through an electron beam at a high power supply when the electron beam is shot via the ultrathin specimen. The electron–crystal interaction produces a TEM image to distinguish the neighboring microstructural features of the material, called resolution.

20.2 Principle

Similar to scanning electron microscopy (SEM), we can use the electron gun via thermionic emission or field electron emission processes to produce the probe beam for TEM. Field electron emission relies on the tunneling effect when we use a strong electric field to induce electrons at the surface of the tip. Thermionic emission is based on the heat effect when we rise the temperature of the filament till the emission of electrons at the surface of the filament.

Optical Properties of Solids: An Introductory Textbook
Kitsakorn Locharoenrat
Copyright © 2016 Pan Stanford Publishing Pte. Ltd.
ISBN 978-981-4669-06-1 (Hardcover), 978-981-4669-07-8 (eBook)
www.panstanford.com

The obtained current density J can be calculated from Richardson's formula as

$$J = RT^2 e^{-\Phi/kT} \qquad \left[\frac{A}{m^2}\right]$$ (20.1)

where T is the temperature of the filament, Φ is the work function, and k is the Boltzmann constant.

Richardson's constant is

$$R = \frac{4\pi\, m\, q\, k^2}{h^3} = 1.20 \times 10^6 \qquad \left[\frac{A}{(mT)^2}\right]$$

where m is the electron mass, q is the electric charge, and h is Planck's constant.

After the diverging beam from the electron gun is released by an accelerating voltage, the beam flows to the anode aperture and the electromagnetic condensed lens, respectively, to compress the beam spot before it passes through the thin sample. Here we can determine the force $\vec{F}$ (in spiral motion) acting on the charged electrons as

$$\vec{F} = q\vec{v} \times \vec{B} = qvB\sin\theta$$ (20.2)

where q is the electric charge, $\vec{v}$ is the beam velocity, $\vec{B}$ is the magnetic field intensity, and θ is the angle between $\vec{v}$ and $\vec{B}$.

After passing through the condensed lens, the beam passes through the well-prepared specimen. The illuminated beam is then directed onto the objective lens before the electron distribution sends the TEM image to the CCD camera. By manipulating the optical aperture behind the objective lens at the appropriate position, the diffraction beam (dark-field image), transmission beam (bright-field image), and diffraction pattern are then created with the high resolution power. These beams offer information about electron density, phase, and periodicity.

The image resolution power r is related to the wavelength λ of the beam as

$$r = \frac{0.612\,\lambda}{n\sin\theta} = \frac{0.612\,\lambda}{NA}$$ (20.3)

where n is the medium refractive index, θ is (1/2) radiation angle from the front of the objective lens, and NA is the numerical aperture.

For electron rest mass m_0 and velocity v accelerated by the high voltage V, the electron wavelength λ is modified according to the relativistic effect as

$$\lambda = \frac{h}{\sqrt{2m_0 eV\left[1 + \dfrac{eV}{2m_0 c^2}\right]}} \tag{20.4}$$

where h is Planck's constant, c is the light velocity, and e is the electric charge.

If we substitute Eq. 20.4 in Eq. 20.3, we finally get

$$r = \frac{0.612\,\lambda}{NA} \frac{h}{\sqrt{2m_0 eV\left[1 + \dfrac{eV}{2m_0 c^2}\right]}} \tag{20.5}$$

The acceleration voltage is used to classify the types of TEM since the resolution of the TEM image depends on it, as shown in Table 20.1.

Table 20.1 Various types of TEM according to the acceleration voltage

Types of TEM	Power supply
Low-voltage TEM	100–200 kV
Medium-voltage TEM	200–500 kV
High-voltage TEM	500 kV to 1 MV

Figure 20.1 shows a model of the diffraction pattern from a single crystal. The crystal plane $\{h, k, l\}$ in the diffraction pattern can be calculated via the camera's formula as

$$d = \frac{\lambda L}{R} \tag{20.6}$$

where d is the gap of the crystal planes, λ is the wavelength of the electron, L is the camera length, and R is the gap between a bright spot at the middle of the diffraction pattern and a bright spot of interest (black arrow in Fig. 20.1).

If a lattice parameter $a = d\sqrt{h^2 + k^2 + l^2}$ (e.g., a_{cu} = 3.61 Å, a_{Pt} = 3.92 Å), we can obtain $N = h^2 + k^2 + l^2$. We finally achieve the crystal plane $\{h, k, l\}$, as shown in Table 20.2.

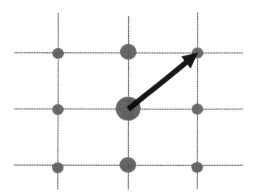

Figure 20.1 Model of the electron diffraction pattern from a single crystal.

Table 20.2 Some possible $N = h^2 + k^2 + l^2$ values compared with the crystal plane

$N = h^2 + k^2 + l^2$	$\{h, k, l\}$	$N = h^2 + k^2 + l^2$	$\{h, k, l\}$
1	100	15	—
2	110	16	400
3	111	17	410
4	200	18	411 or 330
5	210	19	331
6	211	20	420
7	—	21	421
8	220	22	332
9	221 or 300	23	—
10	310	24	422
11	311	25	500
12	222	26	510
13	320	27	511 or 333
14	321		

Finally, the TEM image is then directed onto a fluorescent screen and magnified by changing the gap of the crossover-screen. We can also adjust the current of the lens to vary the focal length. This results in the image resolution.

20.3   Sample Preparation

The brittle samples with a well-defined cleavage plane (e.g., Si, GaAs, NaCl, MgO, etc.) should be thin enough for the electron beam to penetrate. If we cannot use a razor blade, ultramicrotomy is a good tool for slicing the samples. The prepared slices should be of less than 3 mm width and 100 nm thickness before placing them on a grid served as a support material, as shown in Fig. 20.2 (a thin round metal sheet made of copper or titanium with 3 mm diameter), for effective TEM analysis with good resolution. In general, TEM grid often uses in between 100 and 400 mesh.

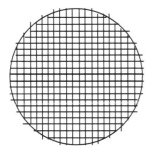 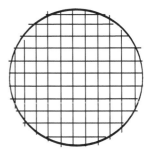

Figure 20.2   Grids served as specimen support.

For ultramicrotomy, the specimen is encapsulated in a thermo-setting polymer (e.g., epoxy resin). This sandwiched specimen is fixed and cleaved over the sharp knife of ultramicrotomy. The thin slices drift off on the surface of the medium (e.g., water). We can then pick up these prepared slices and fit them on a grid.

20.4   Implementation

A TEM system is displayed in Fig. 20.3. The main components of TEM are gun, lens system, image formation, and image recorder.

We first vent the chamber using the nitrogen line. Then we introduce the well-prepared specimen into the chosen holder within the chamber. After that we fill liquid nitrogen into the system and wait for a vacuum of approximately 7×10^{-5} mBar. Next we turn the beam source on (i.e., 10–1000 kV). The electron gun introduces a beam of electrons. This beam hits the condensed lens before going to the specimen. Then the transmitted beam passes through the objective lens to form the TEM image. Finally, the image from the beam is expanded at a proper focal length before appearing on the fluorescent screen and mostly recorded by a CCD camera. At the end, we turn the beam off and vent to remove the specimen.

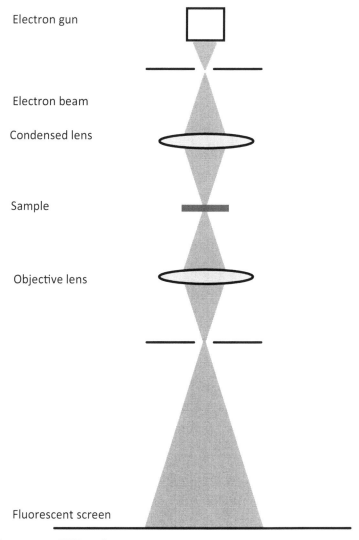

Electron gun

Electron beam

Condensed lens

Sample

Objective lens

Fluorescent screen

Figure 20.3 TEM configuration.

20.4.1 Gun

Similar to SEM, two types of electron guns are used for TEM: thermionic and field emission (Table 20.3). A field emission gun is generally better than a thermionic gun in terms of brightness, monochromatic electron source, and resolution. In thermionic emission, we can use tungsten or lanthanum hexaboride (LaB_6) served as a filament. In terms of performance, as the work function of LaB_6 is lower than that of tungsten, LaB_6 has more ability to emit electrons than tungsten when equal energy is used. If we require a high resolution power, LaB_6 is preferred. This is because since the tip of LaB_6 is smaller than that of tungsten, the electron beam crossover has a smaller diameter than tungsten.

Table 20.3 Three types of electron source operating at 100 kV

Properties	LaB_6	Tungsten	Field emission
Work function (eV)	2.4	4.5	4.5
Temperature (K)	1700	2700	300
Current density (A/m²)	10^6	5×10^4	10^{10}
Crossover size (μm)	10	50	< 0.01
Brightness (A/m² sr)	5×10^{10}	10^9	10^{13}
Vacuum (Pa)	10^{-4}	10^{-2}	10^{-8}

20.4.2 Lens System

After the electron source introduces the probe beam, the beam passes through the electromagnetic condensed lens to control the beam diameter. This illuminated beam then focuses on the specimen. The beam path from the specimen strikes the objective lens. The beam is thus magnified on the fluorescent screen. The lens is electromagnetic in nature. It consists of a cylindrical soft metal core (pole piece) with a hole drilled through it (bore) and wound with the copper wire. When the current passes through the coils, a magnetic field is created around the bore. The focal length of the lens can be reduced if we increase the current of the lens.

20.4.3   Image Formation

By manipulating the optical aperture behind the objective lens at the appropriate position, the diffraction beam (dark-field image), transmission beam (bright-field image), and diffraction pattern are then created. These beams offer information about electron density, phase, and periodicity.

20.4.4   Image Recorder

All the focused electrons behind the objective lens are finally detected and imaged by the CCD camera. Figure 20.4 shows an example of TEM images from metallic nanowires.

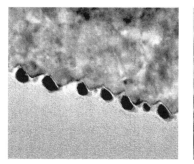 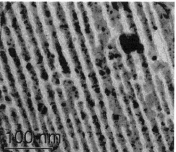

Figure 20.4   TEM images of Pt on MgO (left) and Cu on NaCl (right) substrates.

20.5   Strong Points and Weak Points

Some strong and weak points of TEM are summarized in Table 20.4.

Table 20.4   Strong and weak points of TEM

Strong points	Weak points
Suitable for research and industrial work.	Big equipment and not cheap.
Offers data about the composition and structure of the materials of interest.	Not easy to prepare specimen.
Offers high resolution images as compared to conventional optical microscope.	Specimen should be very thin, tolerate the vacuum conditions, and small enough to fit in the specimen holder.

20.6   Problems

1. Discuss the development of TEM.
2. How to prepare a thin sample?
3. Discuss the differences between thermionic and field emission.
4. Explain the limitations of TEM.
5. Write the TEM operation in a block diagram.
6. Discuss the resolution limit.

Chapter 21

Confocal Microscopy

21.1 Introduction

In this chapter, we will focus on modern confocal microscopy in the mode of sum-frequency generation (SFG) instead of conventional microscopies such as infrared microscopy and Raman microscopy. The optical response from SFG can be remarkably observed in asymmetric materials. The old generation of sum-frequency (SF) microscopy, in principle, works well for a slim (two-dimensional) sample because a whole sample is expected to be on the same focal plane. By contrast, the resolution of a dense sample is poor because the in-focus image can only be detected near the surface of the sample, whereas the out-of-focus image exists in deeper regions of the sample. To overcome this drawback, we can design and construct up-to-date confocal SF microscopy.

21.2 Parameters

We need to understand some parameters or terms of microscopy before heading to confocal SF microscopy, such as view angle, field of view, working distance, magnification, numerical aperture, brightness, resolution, and depth of focus.

Optical Properties of Solids: An Introductory Textbook
Kitsakorn Locharoenrat
Copyright © 2016 Pan Stanford Publishing Pte. Ltd.
ISBN 978-981-4669-06-1 (Hardcover), 978-981-4669-07-8 (eBook)
www.panstanford.com

21.2.1 Angle of View

The angle of view (α), as seen in Fig. 21.1, is explained in terms of the length of focal plane (f) and the image size (d) as

$$\alpha = 2\tan^{-1}\frac{d}{2f} \tag{21.1}$$

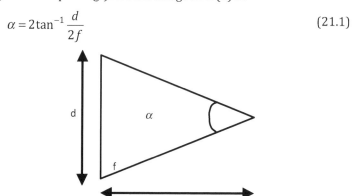

Figure 21.1 Concept of angle of view.

The length of focal plane (f) is calculated by using two parameters—stated focal length (F) and magnification factor (m)—and is written as

$$f = (1 + m)\, F \tag{21.2}$$

The image size (d) can be measured in one of three ways: horizontal, vertical, or diagonal.

21.2.2 Field of View

The field of view (FOV) is written as

$$\text{FOV} = \frac{\text{Eyepiece FN}}{M(\text{ob})}\ [\text{mm}] \tag{21.3}$$

where field number (FN) is the diaphragm diameter seen via eyepiece. It is used to determine the image area of the sample. $M(\text{ob})$ is the magnification of the objective lens or

$$M(\text{ob}) = \frac{f_{\text{tu}}}{f_{\text{ob}}} \tag{21.4}$$

where f_{tu} is the focal length of the tube lens and f_{ob} is the focal length of the objective lens.

21.2.3 Working Distance

After focusing the sample, we can define the working distance as the gap between the front plane of the objective lens and the sample surface, as seen in Fig. 21.2.

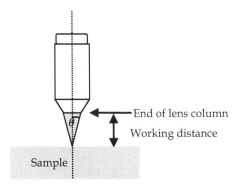

Figure 21.2 Concept of working distance.

21.2.4 Total Magnification

When we observe the image through the binocular, we have the total magnification (M) as

$$M = M(\text{ob}) \times M(\text{oc}) \tag{21.5}$$

where $M(\text{ob})$ and $M(\text{oc})$ are the magnification of the objective lens and the eyepiece, respectively.

21.2.5 Numerical Aperture

The numerical aperture (NA) is written as

$$NA = n \sin \theta \tag{21.6}$$

where n is the refractive index of the surroundings of the objective lenses and the sample (i.e., $n = 1.0$ for air and $n = 1.5$ for oil) and θ is (1/2) radiation angle from the front of the objective lens.

21.2.6 Visual Field Brightness

The visual field brightness (B) is written as

$$B = \frac{(NA)^2}{M(ob)^2} \tag{21.7}$$

The reduced objective magnification with increasing NA enhances brightness.

21.2.7   Resolving Power

The resolving power (r) is written as

$$r = \frac{0.612\lambda}{NA} \quad [\mu m] \tag{21.8}$$

where λ is the wavelength or the radiation in use (i.e., $\lambda = 0.55$ μm in the case of visible light).

High resolving power indicates that the ability of two smallest points can be distinguished.

21.2.8   Depth of Focus

The depth of focus (DOF) is the depth of the sample layer that is still in focus all the time. It is written as

$$DOF = \frac{\omega \times 250000}{NA \times M} + \frac{\lambda}{2(NA)^2} \quad [\mu m] \tag{21.9}$$

where ω is the resolving power of the eyes (i.e., $\omega = 0.0014$ in the case of optical angle = 0.5°). The depth of focus becomes smaller when the NA is larger.

21.3   Principle

In the nonlinear optical phenomena, the nonlinear polarization $\vec{P}$ is written as

$$\vec{P} = \varepsilon_0 [\chi^{(1)}\vec{E} + \chi^{(2)}\vec{E}^2 + \chi^{(3)}\vec{E}^3 + \cdots] \tag{21.10}$$

$$\vec{P} = \vec{P}^{(1)} + \vec{P}^{(2)} + \vec{P}^{(3)} + \cdots \tag{21.11}$$

$$\vec{P} = \vec{P}^{(1)} + \vec{P}^{NL} \tag{21.12}$$

Where $\chi^{(1)}$, $\chi^{(2)}$,... are nonlinear tensor elements, $\vec{P}^{NL}$ is the nonlinear polarization, $\vec{E}$ is the electric field, and ε_0 is permittivity.

For a second-order nonlinear effect, the ith component of nonlinear polarization $P_i^{(2)}$ is written as

$$P_i^{(2)}(\omega_i) = \sum_{jk} \varepsilon_0 \chi_{ijk}^{(2)} E_j(\omega_j) E_k(\omega_k) \qquad (21.13)$$

The frequency mixing phenomenon from Eq. 21.13 can be categorized as sum-frequency generation

$$\omega_3 = \omega_1 + \omega_2 \qquad (21.14)$$

difference-frequency generation

$$\omega_3' = |\omega_1 - \omega_2| \qquad (21.15)$$

and second-harmonic generation

$$\omega_3 = 2\omega_1 \qquad (21.16)$$

From Eq. 21.14, when a non-centrosymmetric medium (e.g., ZnS polycrystal) is irradiated by two coherent lights—one beam at IR light ($\omega_1 = \omega_{IR}$) and another beam at visible light ($\omega_2 = \omega_{VIS}$)—this asymmetric structure then generates a sum-frequency generation ($\omega_3 = \omega_{SFG}$) served as the second-order nonlinear process.

21.4   Implementation

The contract SF images of asymmetric materials (e.g., ZnS pellet with 10 mm ϕ, 3 mm thickness) can be detected according to the setup in Fig. 21.3. The visible light (λ_{VIS} = 532 nm) is produced by the Nd^{3+}:YAG laser. The IR light ($\lambda_{IR} \sim$ 3 μm) is generated by the optical parametric generator (OPG) and the optical parametric amplifier (OPA) systems driven by the same laser. The visible and IR lights are focused on the specimen at a certain angle. After that the medium generates the SF beam (λ_{SF} = 460 nm).

The SF signal passes through the objective lens of a commercial microscope at a certain magnification (say, at ×20). The light passes through the dichroic mirror, band-pass filter, pinhole (e.g., 1 or 0.4 mm ϕ), and photomultiplier in the given order. The SF image is recorded by the charge-coupled device camera.

Figure 21.4a displays the conventional image of ZnS when we illuminate the sample by white light.

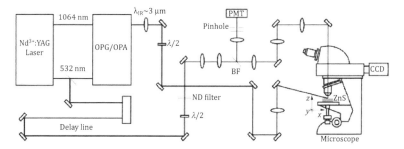

Figure 21.3 Setup of confocal SF microscopy. OPG and OPA are the optical parametric generator and optical parametric amplifier, respectively. PMT, BF, and CCD are photomultiplier, birefringent filter, and charge-coupled device camera, respectively.

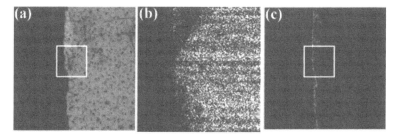

Figure 21.4 (a) Conventional image of ZnS (3 mm thick) when we illuminate it by white light. (b) Confocal SF image of ZnS (1 mm ϕ pinhole). (c) Side illumination image of imperfect ZnS.

Figure 21.4b displays the confocal SF image of ZnS (1 mm ϕ pinhole). Figure 21.4c shows the defective SF image of ZnS in side illumination.

The spatial resolution of confocal SF microscopy is modified from Eq. 21.8 as

$$r = \frac{0.612\,\lambda}{\sqrt{2}\mathrm{NA}} \tag{21.17}$$

where λ is the wavelength in use and NA is the numerical aperture. For instance, if we have $\lambda_{\mathrm{SFG}} = 460$ nm and NA = 0.45, we get $r = 0.4$ µm.

The penetration depth d_{p} of confocal SF microscopy is modified from Eq. 21.9 as

$$d_{\mathrm{p}} = \frac{1.4n\lambda}{\mathrm{NA}^2} \tag{21.18}$$

where n is the medium's refractive index. If we have $n_{\mathrm{ZnS}} = 2.29$, $\lambda_{\mathrm{SFG}} = 460$ nm, and NA = 0.45, we get $Z = 7.3$ μm.

Figure 21.5 shows the optical SF images of ZnS planes in various regions as compared to conventional images.

Figure 21.5a shows the conventional image of ZnS when we illuminate it by white light. Figures 21.5b–d show the different contrasts (a difference of photon density is represented by the white arrow) in the confocal SF images of ZnS (0.4 mm φ pinhole) at 0, 5, and 10 μm depths, respectively, partly due to the nonuniformity of ZnS.

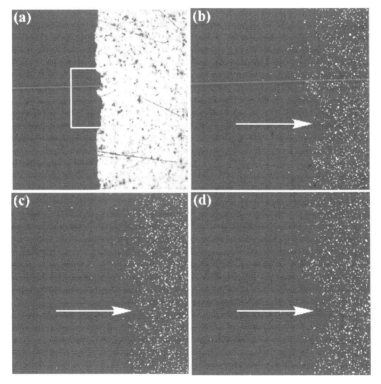

Figure 21.5 (a) Conventional image of ZnS (3 mm thick) when we illuminate it by white light. Confocal SF images of ZnS (0.4 mm φ pinhole) at (b) 0 μm, (c) 5 μm, and (d) 10 μm away from the sample surface.

21.5 Problems

1. Discuss the microscopy parameters.
2. Discuss the resolution limit.
3. Derive the resolutions of conventional confocal microscopy and modern confocal SF microscopy.
4. Derive the depth of focus of conventional confocal microscopy and modern confocal SF microscopy.
5. Explain nonlinear optical effect existing in confocal SF microscopy.
6. Show some examples of SFG in terms of $\omega_3 = \omega_1 + \omega_2$ in non-centrosymmetric media.
7. Discuss the effect of pinhole on the resolution of confocal SF microscopy.
8. Give some examples of confocal SF microscopy applications.

Chapter 22

Tissue Structure

22.1 Introduction

Before heading to optical tomography in turbid media, in this chapter we will first understand the fundamentals of some optical phenomena in biological species, especially in cells and tissues. Tissues can be epithelial, connective, muscle, and nervous. Cells are the smallest structural and functional units of life. Cells include a nucleus, mitochondria, lysosome, endoplasmic reticulum, and golgi.

22.2 Light and Tissue Interaction

Optical signals from tissues provide their structural and functional information. Generally, there are two main types of optical signals. They come from the electronic (absorption) and vibrational (elastic scattering) transitions in the optical tissue absorber, as seen in Fig. 22.1. Biological tissues are turbid, and light is scattered strongly. Thus, the path lengths of light cannot be defined uniquely.

Consider a spherical absorber (i.e., chromophore) at a certain dimension. This absorber is assumed to be of good uniformity. The absorption coefficient μ_a (cm^{-1}) can then be explained in terms of

$$\mu_a = \sigma_a N_a \tag{22.1}$$

Optical Properties of Solids: An Introductory Textbook
Kitsakorn Locharoenrat
Copyright © 2016 Pan Stanford Publishing Pte. Ltd.
ISBN 978-981-4669-06-1 (Hardcover), 978-981-4669-07-8 (eBook)
www.panstanford.com

where σ_a is the absorption cross section (cm^2) and N_a is the number of absorbing molecules per unit volume (cm^{-3}).

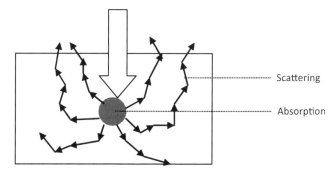

Scattering

Absorption

Figure 22.1 Absorption and scattering processes on a spherical absorber.

The absorption mean free path l_a (m) represents the average distance the photon travels before being absorbed or

$$l_a = \frac{1}{\mu_a} \tag{22.2}$$

The relation between the light intensity absorbed in the slab, $dz(dI)$, and the intensity leaving the tissue (I_0) is written as

$$\frac{dI}{I_0} = -\sigma_a N_a dz \tag{22.3}$$

$$\int_{I_0}^{I} \frac{dI}{I(z)} = -\int_{0}^{b} \sigma_a N_a dz \tag{22.4}$$

$$-\ln\frac{I}{I_0} = \sigma_a N_a b \tag{22.5}$$

Since

$$N_a\left(\frac{\text{molecule}}{\text{cm}^3}\right)\cdot\left(\frac{1\,\text{mol}}{6.023\times10^{23}\,\text{molecules}}\right)\cdot\left(\frac{1000\,\text{cm}^3}{1\,\text{L}}\right) = c\left(\frac{\text{mol}}{\text{L}}\right)$$

and

$$\ln(x) = \frac{\log(x)}{\log(e)} = 2.303\log(x)$$

then Eq. 22.5 is written as

$$-\log\frac{I}{I_0} = \frac{6.023\times10^{20}}{2.303}\sigma_a\, c\, b = \varepsilon\, c\, b \tag{22.6}$$

where ε is the molar extinction coefficient $(cm^{-1}\, M^{-1})$.

If transmittance (T) and absorbance (A) are defined as

$$T = \frac{I}{I_0} = e^{-\varepsilon c b} = 10^{-A} \tag{22.7}$$

we get

$$\ln\frac{I}{I_0} = \ln(10^{-\varepsilon c b}) = -2.303\varepsilon c b \tag{22.8}$$

$$I = I_0 e^{-2.303\varepsilon c b} = I_0 e^{-\mu_a b} \tag{22.9}$$

where $\mu_a = 2.303\varepsilon c$

By measuring the transmittance T and absorbance A of the absorber, as summarized in Table 22.1, we get ε from Eq. 22.7. With the knowledge of ε, if we measure μ_a, we can quantify the concentration of chromophores from Eq. 22.9.

Table 22.1 Major tissue absorbers existing in various wavelengths

UV	VIS-NIR	IR
Amino acid	Cytochrome a3	Glucose
Fatty acid	Hemoglobin (oxy and deoxy)	Protein
Protein	Lipids (β-carotene)	Water
Peptide		
DNA		
NADH, FAD		

On the other hand, when there are inhomogeneities in the bulk, optical scattering exists. For example, Mie and Rayleigh scatterings generally exist in biological tissues. For this fluctuation, the medium refractive index is a nonuniformity. The refractive index mismatch area reduces the light intensity. The scattering coefficient μ_s (cm^{-1}) is then written as

$$\mu_s = \sigma_s\, N_s \tag{22.10}$$

where σ_s is the scattering cross section (cm^2) and N_s is the number of scattering molecules per unit volume (cm^{-3}).

The scattering mean free path l_S (m) represents the average distance the photon travels during the scattering event or

$$l_S = \frac{1}{\mu_S} \qquad (22.11)$$

Decreased scattering coefficient μ_S' and anisotropic constant $g = \langle \cos \theta \rangle$ relate to the scattering coefficient μ_S as

$$\mu_S' = (1-g)\mu_S \qquad (22.12)$$

where g can be -1 (totally backward scattering), 0 (isotropic scattering), or 1 (totally forward scattering). In general, biological tissues have this value in the range of 0.65 to 0.95.

In practice, optical absorption and scattering always appear at the same time. They rely on the wavelength as

$$\mu_S' = \frac{A}{\lambda^b} \qquad (22.13)$$

From Eq. 22.13, both scattering and absorption are mostly suppressed, whereas absorption is dominant if the wavelength is increased from UV to IR in biological tissues.

Light–tissue interactions can be modeled in many ways. When $\mu_a \gg \mu_S'$, we use the Lambert–Beer law ($\lambda \leq 300$; $\lambda \geq 2000$ nm). In the case of $\mu_S' \gg \mu_a$, we use the diffusion approximation ($600 \leq \lambda \leq 1000$ nm). If $\mu_S' \approx \mu_a$, we can use the Monte Carlo method ($300 \leq \lambda \leq 600$ nm; $1000 \leq \lambda \leq 2000$ nm). Light transport in tissues is expected to offer some functional data about tissue, such as O_2 consumption and hemodynamics. Due to the therapeutic window ($600 \leq \lambda \leq 1000$ nm), the diffusion theory through the time-resolved method is suggested to carry more information about the path length of light. This theory is, therefore, very useful to the quantification of drug delivery.

The time-dependent optical diffusion equation can be written as

$$\left[\frac{1}{c}\frac{\partial}{\partial t} - \nabla \cdot D(r)\nabla + \mu_a(r) \right] \Phi(r,t) = S(r,t) \qquad (22.14)$$

for the analysis

where $D(r) = \dfrac{1}{3\left[\mu_a(r) + \mu_S(r) \right]}$

and c is the light velocity, D is the diffusion coefficient, Φ is the light flux, S is the source term, r is the measuring position, and t is the measuring time.

The image reconstructions of the scattering and absorption coefficients are called diffuse optical tomography (DOT). Usually, solving this equation is a difficult task. Simple fluorescence diffuse optical tomography (FDOT) algorithm, which is thus reduced to a DOT problem, successfully reconstructs the image of the fluorescence target. The concept of virtual total light is very useful, and it will be discussed in Chapter 24.

22.3   Problems

1. What is tissue?
2. Discuss the absorption and scattering processes and their parameters.
3. Give some examples of absorbers and scatterers in tissues.
4. Explain the types of optical signal.
5. Explain the metrics of absorption.
6. Discuss the Lambert–Beer law in absorption fundamental.
7. Discuss the difference between elastic and inelastic scattering.
8. Explain the concepts of Rayleigh and Mie scatterings.
9. Write the diagram of hierarchy of ultrastructure in tissue.
10. Compare the modeling method to understand light propagation in tissue.

Chapter 23

Optical Coherence Tomography

23.1 Introduction

Tomography is imaging the interior structure of any matter without invasion to identify the locations and profiles of the matter of interest. Some tomographic techniques are summarized in Table 23.1.

Table 23.1 Tomographic techniques

Techniques	Resolution	Penetration depth	Tracer
Computed tomography	$\sim$mm	>10 cm	Yes
Magnetic resonance imaging	$\sim$mm	>10 cm	Yes
Positron emission tomography	$\sim$mm	>10 cm	Yes
Confocal microscopy	0.1–1 μm	0.1–1 mm	Yes
Multiphotonmicroscopy	0.1–1 μm	0.1–1 mm	Yes
Ultrasound imaging	$\sim$100 μm	0.1–10 cm	No
Optical coherence tomography	1–15 μm	Several mm	No

In this chapter, we will focus only on optical coherence tomography (OCT). In OCT, we detect the cross-correlation of the path length between the reference arm l_R and the sample arm l_S. An interference signal (using a low-coherence light source) exists if the

Optical Properties of Solids: An Introductory Textbook
Kitsakorn Locharoenrat
Copyright © 2016 Pan Stanford Publishing Pte. Ltd.
ISBN 978-981-4669-06-1 (Hardcover), 978-981-4669-07-8 (eBook)
www.panstanford.com

path length between the reference and sample arms is tunable in the coherence length Δz in which $|l_R - l_s| < \Delta z$.

23.2 Principle

23.2.1 Fundamentals of Electromagnetic Wave

We start with the concept of electromagnetic wave traveling in the z-direction as

$$E_z(z, t) = E_0 e^{i(\omega t - kz)} \tag{23.1}$$

where E_0 is the electric field amplitude, $\omega = 2\pi\upsilon$ is the angular frequency, $k = \dfrac{2\pi}{\lambda}$ is the propagation vector, f is the wave frequency, and λ is the wavelength. From

$$\lambda\upsilon = \frac{c}{n} \tag{23.2}$$

where c is the light velocity (= 3.0×10^8 ms^{-1}; vacuum) and n is the medium refractive index, we get the intensity of the electromagnetic wave (power per unit area) as

$$I = \frac{c}{n}\left\langle \left|E(z,t)\right|^2 \right\rangle \tag{23.3}$$

$$I = \frac{c}{n}\left\langle E_0 e^{i(\omega t - kz)} E_0 e^{-i(\omega t - kz)} \right\rangle \tag{23.4}$$

$$I = \frac{c}{n}\left\langle E_0^2 \right\rangle = \frac{c}{n} E_0^2 \propto E_0^2 \tag{23.5}$$

The relation between the electric field and the spectral function will be as follows: For a *monochromatic wave* with the angular frequency ω_0, the electric field is

$$E_z(z, t) = E_0 e^{i(\omega_0 t - kz)} \tag{23.6}$$

Setting the position at $z = 0$, we get

$$E(t) = E_0 e^{i\omega_0 t} \tag{23.7}$$

Equation 23.7 corresponds to the spectral function $S(\omega)$ as

$$S(\omega) = \int_{-\infty}^{\infty} E(t) e^{-i\omega t}\, dt \tag{23.8}$$

$$S(\omega) = \int\limits_{-\infty}^{\infty} E_0 e^{i\omega_0 t}\, e^{-i\omega t}\, dt \qquad (23.9)$$

$$S(\omega) = E_0 \int\limits_{-\infty}^{\infty} e^{-i(\omega-\omega_0)t}\, dt \qquad (23.10)$$

Due to

$$\delta(\omega - \omega') = \frac{1}{2\pi} \int\limits_{-\infty}^{\infty} e^{i(\omega-\omega')t}\, dt = \frac{1}{2\pi} \int\limits_{-\infty}^{\infty} e^{-i(\omega-\omega')t}\, dt \qquad (23.11)$$

we get

$$S(\omega) = 2\pi E_0 \delta(\omega - \omega_0) \qquad (23.12)$$

If $S_0(\omega) = 2\pi E_0(\omega)$ corresponds to the amplitude of the spectral component with the angular frequency ω, then

$$S(\omega) = S_0 \delta(\omega - \omega_0) \qquad (23.13)$$

For a *multiple wavelength wave*, the electric field is

$$E(t) = \sum_{\omega} E_0(\omega) e^{i\omega t} \qquad (23.14)$$

$$E(t) = \frac{1}{2\pi} \sum_{\omega} S_0(\omega) e^{i\omega t} \qquad (23.15)$$

In the case of a *continuous-wavelength wave*, the electric field is modified as

$$E(t) = \frac{1}{2\pi} \int\limits_{-\infty}^{\infty} S(\omega) e^{i\omega t}\, d\omega \qquad (23.16)$$

Equation 23.16 also corresponds to the spectral function $S(\omega)$ as

$$S(\omega) = \int\limits_{-\infty}^{\infty} E(t) e^{-i\omega t}\, dt \qquad (23.17)$$

$$S(\omega) = \int\limits_{-\infty}^{\infty} \frac{1}{2\pi} \int\limits_{-\infty}^{\infty} S(\omega') e^{i\omega' t}\, d\omega'\, e^{-i\omega t}\, dt \qquad (23.18)$$

$$S(\omega) = \frac{1}{2\pi} \int\limits_{-\infty}^{\infty} \int\limits_{-\infty}^{\infty} S(\omega') e^{i(\omega'-\omega)t}\, dt\, d\omega' \qquad (23.19)$$

Due to

$$\delta(\omega - \omega') = \frac{1}{2\pi} \int_{-\infty}^{\infty} e^{i(\omega - \omega')t} \, dt = \frac{1}{2\pi} \int_{-\infty}^{\infty} e^{-i(\omega - \omega')t} \, dt \qquad (23.20)$$

we get

$$S(\omega) = \int_{-\infty}^{\infty} S(\omega') \, \delta(\omega' - \omega) \, d\omega' \qquad (23.21)$$

23.2.2 Interference of Electromagnetic Wave

The interference signal can be experimentally performed by a *Michelson interferometer*, as shown in Fig. 23.1.

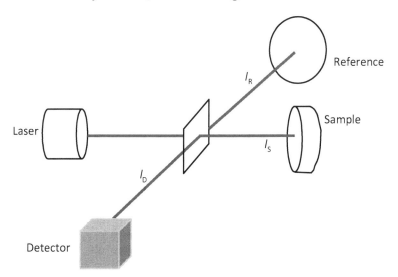

Figure 23.1 Michelson interferometer. l_R, l_S, l_D are the path lengths of the reference arm, sample arm, and detector, respectively.

The measuring time required for a *sample beam* to travel from the beam splitter to the detector is defined as

$$\tau_S = \frac{2l_S + l_D}{c} \qquad (23.22)$$

Where l_R, l_S, l_D are the path lengths of the reference arm, sample arm, and detector, respectively.

The measuring time required for a *reference beam* to travel from the beam splitter to the detector is defined as

$$\tau_R \equiv \tau_S + \tau = \frac{2l_R + l_D}{c} \tag{23.23}$$

where the time delay is defined as

$$\tau = \frac{2(l_R - l_S)}{c} \tag{23.24}$$

The intensity of the electromagnetic wave that impinges on the photodetector is written as

$$I_D(\tau) \propto \left\langle \left| E_D(\tau) \right|^2 \right\rangle \tag{23.25}$$

$$I_D(\tau) = \left\langle \left| E_S(t) + E_R(t - \tau) \right|^2 \right\rangle \tag{23.26}$$

$$I_D(\tau) = \left\langle \left| E_S(t) \right|^2 + \left| E_R(t - \tau) \right|^2 + E_S(t)E_R^*(t - \tau) + E_S^*(t)E_R(t - \tau) \right\rangle \tag{23.27}$$

$$I_D(\tau) = \left\langle \left| E_S(t) \right|^2 \right\rangle + \left\langle \left| E_R(t - \tau) \right|^2 \right\rangle + \left\langle 2\mathrm{Re}\, E_S(t)E_R^*(t - \tau) \right\rangle \tag{23.28}$$

For a coherence (monochromatic) light source, the electric field is written as

$$E(t) = E_0 e^{i\omega_0 t} \tag{23.29}$$

$$E_S(t) = E_{S0} e^{i\omega_0 t} \tag{23.30}$$

$$E_R(t - \tau) = E_{R0} e^{i\omega_0(t - \tau)} \tag{23.31}$$

From the first term in Eq. 23.28, we get

$$\left\langle \left| E_S(t) \right|^2 \right\rangle = \left\langle E_{S0} e^{i\omega_0 t} E_{S0} e^{-i\omega_0 t} \right\rangle = \left\langle E_{S0}^2 \right\rangle = E_{S0}^2 \tag{23.32}$$

From the second term in Eq. 23.28, we get

$$\left\langle \left| E_R(t - \tau) \right|^2 \right\rangle = \left\langle E_{R0} e^{i\omega_0(t - \tau)} E_{R0} e^{-i\omega_0(t - \tau)} \right\rangle = \left\langle E_{R0}^2 \right\rangle = E_{R0}^2 \tag{23.33}$$

From the third term in Eq. 23.28, we get

$$\left\langle 2\mathrm{Re}\, E_S(t)E_R^*(t - \tau) \right\rangle = \left\langle 2\mathrm{Re}\, E_{S0} e^{i\omega_0 t} E_{R0} e^{-i\omega_0(t - \tau)} \right\rangle$$

$$= \left\langle 2\mathrm{Re}\, E_{S0} E_{R0} e^{i\omega_0 \tau} \right\rangle = \left\langle 2\mathrm{Re}\, E_{S0} E_{R0} \cos(\omega_0 \tau) \right\rangle$$

$$= 2\mathrm{Re}\, E_{S0} E_{R0} \cos(\omega_0 \tau) \tag{23.34}$$

Substituting Eqs. 23.32 and 23.34 in Eq. 23.28, we get

$$I_D(\tau) = E_{S0}^2 + E_{R0}^2 + 2\,\mathrm{Re}\,E_{S0}E_{R0}\cos(\omega_0\tau) \tag{23.35}$$

We can plot the intensity with respect to the time delay as shown in Fig. 23.2.

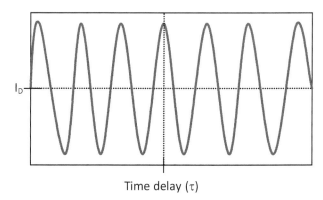

Time delay (τ)

Figure 23.2 Intensity versus time delay from a coherent source.

At the vertical dash line in Fig. 23.2, we have

$$\omega_0\tau = 2\pi \tag{23.36}$$

where

$$\tau = \frac{2(l_R - l_S)}{c} = \frac{2\pi}{\omega_0} \tag{23.37}$$

$$l_R - l_S = \frac{\tau c}{2} = \frac{c\pi}{\omega_0} = \frac{\lambda_0}{2} \tag{23.38}$$

For a low-coherence (continuous-wavelength) light source, the electric field is modified as

$$E(t) = \frac{1}{2\pi}\int_{-\infty}^{\infty} S(\omega)e^{i\phi(\omega)}e^{i\omega t}\,d\omega \tag{23.39}$$

where $\phi(\omega)$ is the frequency-dependent phase due to the optical components in the optical path.

$$E_S(t) = \frac{1}{2\pi}\int_{-\infty}^{\infty} S_S(\omega)e^{i\phi_S(\omega)}e^{i\omega t}\,d\omega \tag{23.40}$$

$$E_R(t-\tau) = \frac{1}{2\pi} \int\limits_{-\infty}^{\infty} S_R(\omega) e^{i\phi_R(\omega)} e^{i\omega(t-\tau)} d\omega \tag{23.41}$$

From

$$I_D(\tau) = \left\langle |E_S(t)|^2 \right\rangle + \left\langle |E_R(t-\tau)|^2 \right\rangle + \left\langle 2\mathrm{Re}\,E_S(t)E_R^*(t-\tau) \right\rangle$$

(which is Eq. 23.28), the first term gives

$$\left\langle |E_S(t)|^2 \right\rangle = \left\langle \frac{1}{(2\pi)^2} \int\limits_{-\infty}^{\infty} \int\limits_{-\infty}^{\infty} S_S(\omega) e^{i\phi_S(\omega)} e^{i\omega t} S_S(\omega') e^{-i\phi_S(\omega')} e^{-i\omega' t} d\omega\, d\omega' \right\rangle \tag{23.42}$$

$$\left\langle |E_S(t)|^2 \right\rangle = \left\langle \frac{1}{(2\pi)^2} \int\limits_{-\infty}^{\infty} \int\limits_{-\infty}^{\infty} \int\limits_{-\infty}^{\infty} S_S(\omega) S_S(\omega') e^{i[\phi_S(\omega)-\phi_S(\omega')]} e^{i(\omega-\omega')t} d\omega\, d\omega'\, dt \right\rangle \tag{23.43}$$

$$\left\langle |E_S(t)|^2 \right\rangle = \left\langle \frac{1}{2\pi} \int\limits_{-\infty}^{\infty} \int\limits_{-\infty}^{\infty} S_S(\omega) S_S(\omega') e^{i[\phi_S(\omega)-\phi_S(\omega')]} \delta(\omega-\omega') d\omega\, d\omega' \right\rangle \tag{23.44}$$

$$\left\langle |E_S(t)|^2 \right\rangle = \left\langle \frac{1}{2\pi} \int\limits_{-\infty}^{\infty} S_S^2(\omega) d\omega \right\rangle \equiv I_S \tag{23.45}$$

The second term in Eq. 23.28 also gives

$$\left\langle |E_R(t-\tau)|^2 \right\rangle = \left\langle \frac{1}{2\pi} \int\limits_{-\infty}^{\infty} S_R^2(\omega) d\omega \right\rangle \equiv I_R \tag{23.46}$$

The third term in Eq. 23.28 gives

$$\left\langle 2\mathrm{Re}\,E_S(t)E_R^*(t-\tau) \right\rangle$$

$$= 2\mathrm{Re}\frac{1}{(2\pi)^2} \int\limits_{-\infty}^{\infty} \int\limits_{-\infty}^{\infty} S_S(\omega) e^{i\phi_S(\omega)} e^{i\omega t} S_R(\omega') e^{-i\phi_R(\omega')} e^{-i\omega'(t-\tau)} d\omega\, d\omega' \tag{23.47}$$

$$= 2\mathrm{Re}\frac{1}{(2\pi)^2} \int\limits_{-\infty}^{\infty} \int\limits_{-\infty}^{\infty} \int\limits_{-\infty}^{\infty} S_S(\omega)S_R(\omega')e^{i[\phi_S(\omega)-\phi_R(\omega')]}e^{i(\omega-\omega')t}e^{i\omega'\tau}d\omega d\omega' dt$$

(23.48)

$$= \frac{2}{2\pi}\mathrm{Re} \int\limits_{-\infty}^{\infty} \int\limits_{-\infty}^{\infty} S_S(\omega)S_R(\omega')e^{i[\phi_S(\omega)-\phi_R(\omega')]}e^{i\omega'\tau}\delta(\omega-\omega')d\omega d\omega'$$

(23.49)

$$= \frac{1}{\pi}\mathrm{Re} \int\limits_{-\infty}^{\infty} S_S(\omega)S_R(\omega)e^{i[\phi_S(\omega)-\phi_R(\omega)]}e^{i\omega\tau}d\omega$$

(23.50)

$$= \frac{1}{\pi}\mathrm{Re} \int\limits_{-\infty}^{\infty} S_S(\omega)S_R(\omega)e^{i\Delta\phi(\omega)}e^{i\omega\tau}d\omega$$

(23.51)

Substituting Eqs. 23.45, 23.46, and 23.51 in Eq. 23.28, we get

$$I_D(\tau) = I_S + I_R + \frac{1}{\pi}\mathrm{Re} \int\limits_{-\infty}^{\infty} S_S(\omega)S_R(\omega)e^{i\Delta\varphi(\omega)}e^{i\omega\tau}d\omega$$

(23.52)

Considering the light source with a Gaussian spectral distribution, we have

$$\Delta\phi(\omega) = 0$$

(23.53)

$$S_S(\omega) = S_{S0}e^{-\frac{2\ln 2(\omega-\omega_0)2}{\Delta\omega^2}}$$

(23.54)

$$S_R(\omega) = S_{R0}e^{-\frac{2\ln 2(\omega-\omega_0)2}{\Delta\omega^2}}$$

(23.55)

Substituting Eqs. 23.53–23.55 in Eq. 23.52, we get

$$I_S + I_R + \frac{1}{\pi}\mathrm{Re} \int\limits_{-\infty}^{\infty} S_S(\omega)S_R(\omega)e^{i\Delta\phi(\omega)}e^{i\omega\tau}d\omega$$

(23.56)

$$= I_S + I_R + \frac{1}{\pi}\mathrm{Re} \int\limits_{-\infty}^{\infty} S_{S0}e^{-\frac{2\ln 2(\omega-\omega_0)2}{\Delta\omega^2}} S_{R0}e^{-\frac{2\ln 2(\omega-\omega_0)2}{\Delta\omega^2}} e^{i\Delta\phi(\omega)}e^{i\omega\tau}d\omega$$

(23.57)

$$= I_S + I_R + \frac{1}{\pi}\mathrm{Re} \int\limits_{-\infty}^{\infty} S_{S0}e^{-\frac{2\ln 2(\omega-\omega_0)}{\Delta\omega^2}} S_{R0}e^{-\frac{2\ln 2(\omega-\omega_0)2}{\Delta\omega^2}} e^{i\omega\tau}d\omega \quad (23.58)$$

$$= I_S + I_R + \frac{S_{S0}S_{R0}}{\pi} \mathrm{Re} \int_{-\infty}^{\infty} e^{-\frac{4\ln 2(\omega-\omega_0)^2}{\Delta\omega^2} + i\omega\tau} d\omega \qquad (23.59)$$

$$= I_S + I_R + \frac{S_{S0}S_{R0}}{\pi} \mathrm{Re} \frac{\Delta\omega}{2} \sqrt{\frac{\pi}{\ln 2}} e^{\frac{-\Delta\omega^2\tau^2}{16\ln 2}} e^{i\omega_0\tau} \qquad (23.60)$$

$$= I_S + I_R + \frac{S_{S0}S_{R0}\Delta\omega}{2\sqrt{\pi\ln 2}} e^{\frac{-\Delta\omega^2\tau^2}{16\ln 2}} \cos(\omega_0\tau) \qquad (23.61)$$

$$= I_S + I_R + I_1 e^{-\frac{4\ln 2\tau^2}{\Delta\tau^2}} \cos(\omega_0\tau) \qquad (23.62)$$

where

$$I_1 = \frac{S_{S0}S_{R0}\Delta\omega}{2\sqrt{\pi\ln 2}} \qquad (23.63)$$

and

$$\Delta\tau \equiv \frac{8\ln 2}{\Delta\omega} \qquad (23.64)$$

where

$$\tau = \frac{2(l_R - l_S)}{c} \qquad (23.65)$$

We can then plot the interference signal of the single-layer structure as shown in Fig. 23.3.

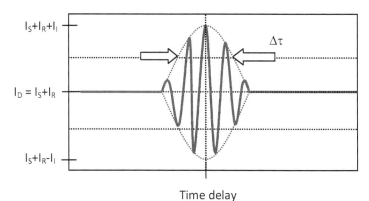

Figure 23.3 Intensity versus time delay for a low coherent source (single-layer structure).

The interference signal of a multi-layer structure with N inner interfaces is

$$I_D(\tau) \propto \left\langle \left| E_D(\tau) \right|^2 \right\rangle \tag{23.66}$$

$$I_D(\tau) = \left\langle \left| \left[\sum_{i=0}^{N} E_{Si}(t - \tau_i) \right] + E_R(t - \tau) \right|^2 \right\rangle \tag{23.67}$$

where

$$\tau = \frac{2(l_R - l_{SO})}{c} \tag{23.68}$$

$$\tau_i = \frac{2(l_{Si} - l_{SO})}{c} \tag{23.69}$$

SO denotes the surface of the sample, and $l_{Si} - l_{SO}$ is the distance between the ith inner interface and the surface of the sample.

$$I_D(\tau) = \left\langle \left| \left[\sum_{i=0}^{N} E_{Si}(t - \tau_i) \right] \right|^2 \right\rangle + \left\langle \left| E_R(t - \tau) \right|^2 \right\rangle + $$

$$\left\langle 2\mathrm{Re} \sum_{i=0}^{N} E_{Si}(t - \tau_i) E_R^*(t - \tau) \right\rangle \tag{23.70}$$

$$I_D(\tau) = I_S + I_R + \sum_{i=0}^{N} I_{Ii} e^{-\frac{4\ln 2(\tau - \tau_i)^2}{\Delta \tau^2}} \cos[(\omega_0(\tau - \tau_i)] \tag{23.71}$$

where

$$I_{Ii} = \frac{S_{Si0} S_{R0} \Delta \omega}{2\sqrt{\pi \ln 2}} \tag{23.72}$$

We can plot the interference signal of a multi-layer structure as shown in Fig. 23.4.

For example, when

$$\tau = \tau_0 = 0 \tag{23.73}$$

we get $l_R = l_{SO}$ (23.74)

As $\tau = \tau_1 = \dfrac{2(l_{S1} - l_{SO})}{c}$ (23.75)

we get $\quad l_R = l_{S1} = l_{SO} + \dfrac{c\tau_1}{2}$  (23.76)

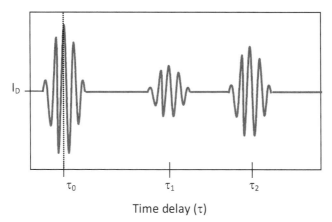

Time delay (τ)

Figure 23.4 Intensity versus time delay for a low coherent source (multiple-layer structures).

23.2.3 Resolution Limit

If we start from the spectral distribution of Gaussian beam optics as

$$S(\omega) = S_0 e^{-\frac{2\ln 2(\omega-\omega_0)^2}{\Delta\omega^2}}$$ (23.77)

we have the power spectral density as

$$\left|S(\omega)\right|^2 = S_0^2 e^{-\frac{4\ln 2(\omega-\omega_0)^2}{\Delta\omega^2}}$$ (23.78)

When $\left|\omega - \omega_0\right| = \dfrac{\Delta\omega}{2}$, we get (Fig. 23.5)

$$\left|S\left(\omega_0 \pm \frac{\Delta\omega}{2}\right)\right|^2 = S_0^2 e^{-\frac{4\ln 2\Delta\omega^2/4}{\Delta\omega^2}} = S_0^2 e^{-\ln 2} = \frac{1}{2}S_0^2$$ (23.79)

When $\left|\tau\right| = \dfrac{\Delta\tau}{2}$, Eq. 23.62 is modified as

$$I_D(\tau) = I_S + I_R + I_1 e^{-\frac{4\ln 2\Delta\tau^2/4}{\Delta\tau^2}} \cos\left(\omega_0 \frac{\Delta\tau}{2}\right)$$ (23.80)

$$I_D(\tau) = I_S + I_R + I_1 e^{-\ln 2} \cos\left(\omega_0 \frac{\Delta\tau}{2}\right) \tag{23.81}$$

$$I_D(\tau) = I_S + I_R + \frac{1}{2} I_1 \cos\left(\omega_0 \frac{\Delta\tau}{2}\right) \tag{23.82}$$

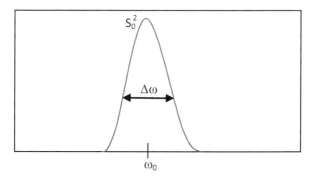

Figure 23.5 Power spectral density versus angular frequency.

If the angular frequency is

$$\omega = \frac{2\pi c}{\lambda} \tag{23.83}$$

then $$\quad \Delta\omega = \frac{2\pi c}{\lambda^2} \Delta\lambda \tag{23.84}$$

From $$\quad \Delta\tau \equiv \frac{8\ln 2}{\Delta\omega} \tag{23.64}$$

and $$\quad l_R - l_S = \frac{c\tau}{2} \tag{23.65}$$

we get

$$\Delta\tau \equiv \frac{4\ln 2}{\pi c}\left(\frac{\lambda^2}{\Delta\lambda}\right) \tag{23.85}$$

We finally have the coherence length Δz or the *axial resolution* as (Fig. 23.6)

$$l_R - l_S = \Delta z = \frac{2\ln 2}{\pi}\left(\frac{\lambda^2}{\Delta\lambda}\right) \tag{23.86}$$

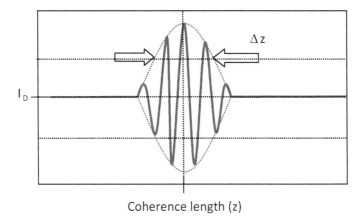

Figure 23.6 Intensity versus coherence length.

From Eq. 23.86, the emission bandwidth of the optical beam will determine the axial resolution. That is, we get better axial resolution (and also contrast images) when the bandwidth of the beam source is broaden.

For a *transverse resolution*, we start from Gaussian beam optics as shown in Fig. 23.7. The electric field variation in the transverse direction is given by

$$E_T = E_{T0}e^{-\frac{r^2}{w^2}} \tag{23.87}$$

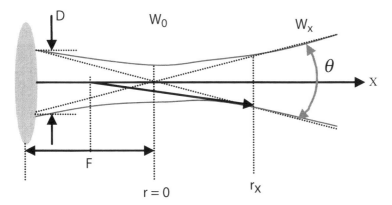

Figure 23.7 Gaussian beam optics. *D* is the point source diameter illuminated along a lens, and *F* is the focal length of the lens.

If w_0 is the radius of the Gaussian wave at position $x = 0$, the radius of the Gaussian wave is written as

$$w_X^2 = w_0^2 \left[1 + \left(\frac{\lambda x}{\pi w_0^2} \right)^2 \right]$$

(23.88)

At a large distance from the beam waist, the angle θ of the diverging beam is given by

$$\theta = \frac{4\lambda}{2\pi w_0} \approx \frac{D}{F}$$

(23.89)

We finally have the transverse resolution as

$$2w_0 = \left(\frac{4\lambda}{\pi} \right) \left(\frac{F}{D} \right)$$

(23.90)

23.3 Implementation

The configuration of an OCT system is given in Fig. 23.8. There are five main components (including the beam splitter).

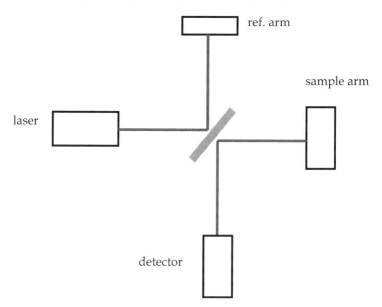

Figure 23.8 Block diagram of OCT configuration.

23.3.1 Optical Sources

A coherence beam is required and summarized in Table 23.2. First, we need the emission of light in NIR, deep in biological tissues (Fig. 23.9). Second, we need a short temporal coherence length. We get better axial resolution (and also contrast images) when the bandwidth of the beam source is broad. Last, we need high irradiance in the case of the poor back-scattering medium under study.

Table 23.2 Short-coherence sources used in OCT systems

Light source	Center wavelength (nm)	Bandwidth (nm)	Emission power (MW)
Edge emission mode LED	1300, 1550	50–100	20–300 µW
Superluminescent diode	800	20–150	1–10
	1300	40–50	15
Multiple mode QW LED/SLD	800	90	15
	1480	90	5
Optimal synthesis of LED	730	n/a	40 µW
Superluminescent Ti:Al2O3 light source	761	138	40 µW
Laser-pumped fluorescent dye	590	40	9
Ti:Sapphire laser	800	260	n/a
Mode-locked Cr^{4+}: forsterite laser	1280	75	30
Pulsed erbium fiber laser	1375	470	4
Continuum from microstructure fiber	725	325	27
	1115	630	50
Superfluorescent optical fibers: * Er-doped	1550	40	10–100
* Tm-doped	1880	80	7
* Nd/Yb-doped	1060	65	108

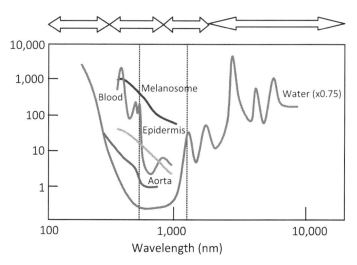

Figure 23.9   Absorption (Abs) versus wavelength in biological tissues.

23.3.2   Interferometers

There are so many types of beam splitters in OCT techniques, and they are summarized in Fig. 23.10.

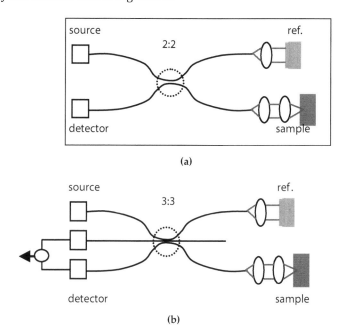

(a)

(b)

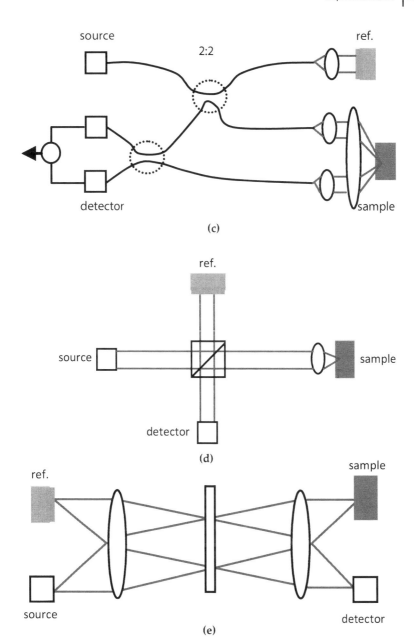

(c)

(d)

(e)

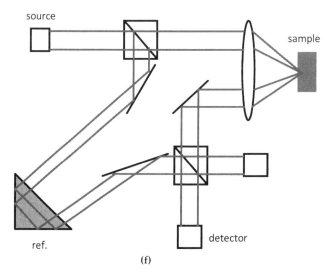

(f)

Figure 23.10 Interferometers used in OCT approaches: (a) Michelson mode in fiberoptic; (b) 3 × 3 fiberoptic coupler mode; (c) two sets of 2 × 2 fiberoptic coupler mode; (d) Michelson mode in free space; (e) an inline mode; and (f) Mach–Zehnder mode in free space.

23.3.3 Reference Arm

In Fig. 23.11, we can use a linear translating mirror. We also can use a fixed mirror with a pair of piezoelectric transducer.

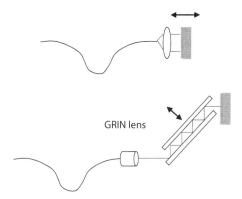

Figure 23.11 Reference optical relay scannings: a linear translating mirror (top) and a fixed mirror with a pair of piezoelectric transducer (bottom).

Furthermore, we can use a fixed mirror with a corner-cube reflector, as shown in Fig. 23.12.

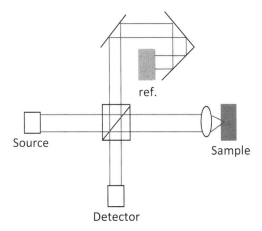

Figure 23.12 A fixed mirror with a corner-cube reflector in the reference arm.

Last we can apply a scanning mirror with reflectometer, as shown in Fig. 23.13.

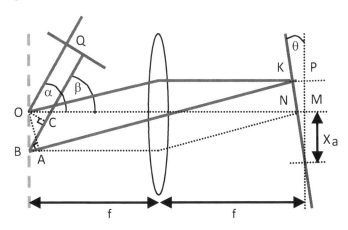

Figure 23.13 Details of a scanning mirror with reflectometer.

In a one-round trip from Fig. 23.13, the phase difference ψ as a function of the spectral term k can be written as

$$\psi(k) = 2k(L_R + OKBC + CQ - L_S - AB - BC) \tag{23.91}$$

$$\psi(k) = 2k(L_R - L_S + CQ - OMO - 2MN - AB) \tag{23.92}$$

But $\delta = L_R - L_S + CQ + OMO$ $\tag{23.93}$

Then $\psi(k) = 2k(\delta - 2MN - AB)$ $\tag{23.94}$

$$\psi(k) = 2k\delta - 4kx_0\theta - 4kf\theta \sin \beta \tag{23.95}$$

If the grating equation is given by

$$p(\sin \alpha + \sin \beta) = m\lambda \tag{23.96}$$

where p is the grating constant, m is the interference order, and λ is the wavelength, and the spectral component at λ_0 travels along the lens as

$$p(\sin \beta = m(\lambda - \lambda_0) \tag{23.97}$$

substituting Eq. 23.97 in Eq. 23.95, we get

$$\psi(k) = 2k\delta - 4kx_0\theta + \frac{8\pi m f \theta(k - k_0)}{pk_0} \tag{23.98}$$

If we consider the phase difference ψ as a function of k_0, we have

$$\psi(k) = \psi(k_0) + \psi'(k_0)(k - k_0) + \psi''(k_0)\frac{(k - k_0)^2}{2!}$$
$$+ \psi'''(k_0)\frac{(k - k_0)^3}{3!} + \cdots \tag{23.99}$$

The phase delay is

$$t_p = \frac{1}{c}\frac{\psi(k_0)}{k_0} = \frac{2\delta}{c} - \frac{4x_0\theta}{c} \tag{23.100}$$

The group delay is

$$t_g = \frac{1}{c}\psi'(k_0) = \frac{L_g}{c} = \frac{2\delta}{c} - \frac{4x_0\theta}{c} + \frac{8\pi m f \theta}{cpk_0} \tag{23.101}$$

where L_g is the group-delay length.

23.3.4 Sample Arm

We can choose the fixed sample on a translating stage or scan with a galvanometer (Fig. 23.14).

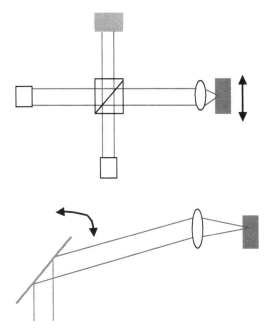

Figure 23.14 Sample arms: a fixed (top) and a moving (bottom) scanner.

23.3.5 Data Acquisition Unit

According to the interference signal of a multi-layer structure for a Gaussian spectrum obtained by the photodetector as (Fig. 23.15)

$$I_D(\tau) = I_S + I_R + \sum_{i=0}^{N} I_{Ii} e^{-\frac{4\ln 2(\tau - \tau_i)^2}{\Delta \tau^2}} \cos[\omega_0(\tau - \tau_i)] \qquad (23.102)$$

where

$$\tau = \frac{2(l_R - l_{SO})}{c} \qquad (23.103)$$

$$\tau_i = \frac{2(l_{Si} - l_{SO})}{c} \qquad (23.104)$$

For example, when

$$\tau = \tau_0 = 0 \qquad (23.105)$$

we get

$$l_R = l_{SO} \qquad (23.106)$$

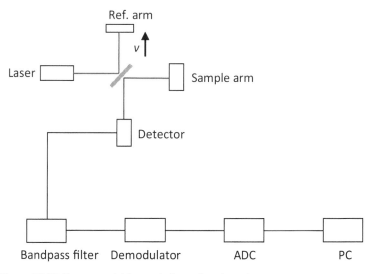

Figure 23.15 Data acquisition unit from the photodetector.

And also as

$$\tau = \tau_1 = \frac{2(l_{Si} - l_{SO})}{c} \tag{23.107}$$

we get

$$l_R = l_{Si} = l_{SO} + \frac{c\tau_1}{2} \tag{23.108}$$

We then finally get

$$t_i = t_0 + \frac{l_{Si} - l_{SO}}{v} = t_0 + \frac{c\tau_i}{2v} \tag{23.109}$$

in which the acquisition time is defined as

$$t \equiv \frac{c\tau}{2v} = \frac{l_R - l_{SO}}{v} \tag{23.110}$$

For instance, $t_0 = \dfrac{c\tau_0}{2v} = 0$ or $t_i = \dfrac{c\tau_i}{2v} = \dfrac{l_{Si} - l_{SO}}{v}$

Substituting Eq. 23.110 in Eq. 23.102, we get

$$I_D(t) = I_S + I_R + \sum_{i=0}^{N} I_{Ii} e^{-\frac{16v^2 \ln 2(t-t_i)^2}{c^2 \Delta \tau^2}} \cos\left[\frac{2v\omega_0}{c}(t - t_i)\right] \tag{23.111}$$

The envelope and the modulation frequency are represented by

$$A_i = I_{1i}e^{-\frac{16v^2 \ln 2(t-t_i)^2}{c^2 \Delta \tau^2}} \quad \text{and} \quad f = \frac{1}{t-t_i} = \frac{1}{2\pi} \cdot \frac{2v\omega_0}{c},$$

respectively (Fig. 23.16).

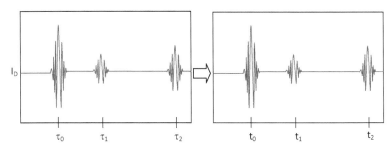

Figure 23.16 Intensity versus time delay (left) and an acquisition time (right).

For example, if the velocity v is 1 mms^{-1}, and the center wavelength λ_0 is 1 μm, we have the modulation frequency as

$$f(t) = \frac{1}{t-t_i} = \frac{1}{2\pi} \cdot \frac{2v\omega_0}{c} = \frac{1}{2\pi} \cdot \frac{2v}{c} \cdot \frac{2\pi c}{\lambda_0}$$

$$f(t) = \frac{1}{t-t_i} = \frac{2v}{\lambda_0} = \frac{2 \times 1\,\text{mms}^{-1}}{1\,\mu\text{m}} = 2\,\text{kHz}$$

For demodulation (Fig. 23.17), we use the lock-in technique and Hilbert transform. In the lock-in technique, the cross-correlation intensity I_{cc} for each interface can be written as

$$I_{CC}(t) = I_{1i}e^{-\frac{16v^2 \ln 2(t-t_i)^2}{c^2 \Delta \tau^2}} \cos\left[\frac{2v\omega_0}{c}(t-t_i)\right] = A_i \cos[\omega_m(t-t_i)]$$

$$(23.112)$$

where $\omega_m = \dfrac{2\pi}{t-t_i} = \dfrac{2v\omega_0}{c}$

Applying the reference source with the frequency ω_r in Eq. 23.112, we get

$$I_{CC}(t) \cos(w_r t) = A_i \cos[\omega_m(t-t_i)] \cos(\omega_r t) \qquad (23.113)$$

$$I_{CC}(t)\cos(\omega_r t) = \frac{A_i}{2}\{\cos[(\omega_r + \omega_m)t - \omega_m t_i)] + \cos[(\omega_r - \omega_m)t + \omega_m t_i)]\}$$

$$(23.114)$$

and we also get

$$I_{CC}(t) \sin(w_r t) = A_i \cos[\omega_m(t - t_i)] \sin(\omega_r t) \tag{23.115}$$

$$I_{CC}(t)\sin(\omega_r t) = \frac{A_i}{2}\{\sin[(\omega_r + \omega_m)t - \omega_m t_i)] + \sin[(\omega_r - \omega_m)t + \omega_m t_i)]\} \tag{23.116}$$

If attenuating the sum frequency component using a low-pass filter is defined as

$$2[I_{CC1}^2(t) + I_{CC2}^2(t)]^{1/2} \tag{23.117}$$

where

$$I_{CC1}(t) = \frac{A_i}{2}\cos[(\omega_r - \omega_m)t + \omega_m t_i] \tag{23.118}$$

$$I_{CC2}(t) = \frac{A_i}{2}\sin[(\omega_r - \omega_m)t + \omega_m t_i] \tag{23.119}$$

we get

$$2[I_{CC1}^2(t) + I_{CC2}^2(t)]^{1/2} = 2\left(\frac{A_i^2}{4}\right)^{1/2} = A_i \tag{23.120}$$

In the Hilbert transform, we have

$$g(t) = \frac{1}{\pi}\int_{-\infty}^{\infty}\frac{f(\tau)}{\tau - t}d\tau = f(\tau) * \frac{-1}{t\pi} \tag{23.121}$$

$$f(t) + ig(t) = Ae^{i\theta}\left\{\begin{array}{l}A \rightarrow \text{envelope} \\ d\theta/dt \rightarrow \text{modulation frequency}\end{array}\right. \tag{23.122}$$

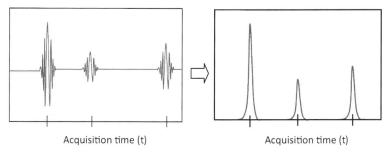

Acquisition time (t)   Acquisition time (t)

Figure 23.17 Before (left) and after demodulation (right).

23.4 Applications

There are so many applications of OCT approaches, and they are summarized in Table 23.3.

Table 23.3 Some OCT applications

OCT methods	Detections
Doppler OCT	Velocity of moving sample
Polarization-sensitive OCT (Fig. 23.18)	Polarization state of reflected photon from sample
Stokes–Mueller OCT	Polarization state from Stokes vector
Optical coherence microscopy	Imaging deeper into a high scatterer
Full-field optical coherence microscopy (Fig. 23.19)	Simultaneous acquisition of all pixels of images
Fourier-domain OCT (Fig. 23.20)	Fourier transformation of interferogram
Spectroscopic OCT (Fig. 23.21)	Morlet wavelet transformation
Time-domain OCT (Fig. 23.22)	Scanning range in mm scale

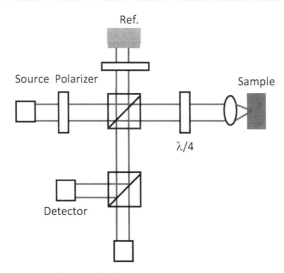

Figure 23.18 Optical setup of polarization-sensitive OCT.

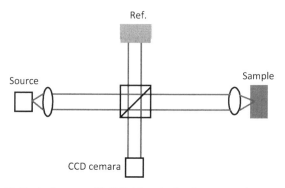

Figure 23.19 Optical setup of full-field optical coherence microscopy.

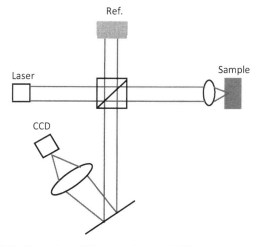

Figure 23.20 Optical setup of Fourier-domain OCT.

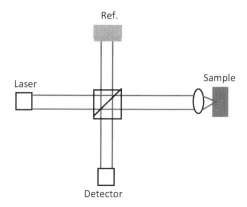

Figure 23.21 Optical setup of spectroscopic OCT.

Figure 23.22 Time-domain OCT system. (Left) Material design. Red line represents light traveling during scanning. (Middle) Retroreflector with mirror arm. (Right) Two-dimensional image of a layer of 11 coverslips.

23.5 Problems

1. Derive the intensity of an electromagnetic wave.
2. Prove that

$$I = \frac{c}{n} \left\langle \left| E(z,t) \right|^2 \right\rangle$$

3. Discuss the resolution between axial and transverse one.
4. Write the configuration of an OCT system.
5. Explain endoscopic OCT.
6. Discuss how to convert from a delay time to an acquisition time in DAQ unit.
7. Discuss terms of modulation and demodulation.
8. Give some examples of OCT methods.

Chapter 24

Fluorescence Diffuse Optical Tomography

24.1   Introduction

In this last chapter, we will concentrate on fluorescence diffuse optical tomography (FDOT) used as the probe in deep biological tissue. A joint differential equation of FDOT (excitation and emission) can be reduced to a problem of two separate differential equations of diffuse optical tomography under the same assumptions. In FDOT, we use a fluorescent dye (e.g., indocyanine green) to increase tissue contrast between normal and disordered ones (e.g., breast cancer tissues). Fluorescence itself offers functional information about a tissue's nature, such as oxygenation.

24.2   Principle

We use a diffusion equation to model light transport to the tissue of interest. The flux $\Phi_{ex}(r, t)$ of excitation light is

$$\left[\frac{1}{c} \frac{\partial}{\partial t} - \nabla \cdot D_{ex}(r) \nabla + \mu_a^{ex}(r) \right] \Phi_{ex}(r,t) = S_{ex}(r,t) \qquad (24.1)$$

where r and t are the measuring position and time, respectively, and D and S are the diffusion coefficient and the source term, respectively.

Optical Properties of Solids: An Introductory Textbook
Kitsakorn Locharoenrat
Copyright © 2016 Pan Stanford Publishing Pte. Ltd.
ISBN 978-981-4669-06-1 (Hardcover), 978-981-4669-07-8 (eBook)
www.panstanford.com

The absorption coefficient $\mu_a^{ex}(r)$ can be separated by two parts: those of the fluorophore, $\mu_{a,fluo}^{ex}(r)$, and the background, $\mu_{a,back}^{ex}(r)$.

If the fluorescence lifetime, quantum efficiency, absorption cross section, and density of fluorophores are denoted by τ, γ, σ, and N, respectively, the fluorescence emission Φ_{em} will be expressed by the flux of the excitation light as

$$\left[\frac{1}{c}\frac{\partial}{\partial t} - \nabla \cdot D_{em}(r)\nabla + \mu_a^{em}(r)\right]\Phi_{em}(r,t) = \frac{\gamma\sigma N(r)}{\tau}\int_0^t dt'\Phi_{ex}(r,t')e^{-\frac{t-t'}{\tau}}$$

$$(24.2)$$

As the inverse convolution operator L^{-1} and the left-hand side operators can be exchanged, the integral of the right-hand side of Eq. 24.2 is simplified as

$$\left[\frac{1}{c}\frac{\partial}{\partial t} - \nabla \cdot D_{em}(r)\nabla + \mu_a^{em}(r)\right]L^{-1}\Phi_{em}(r,t) = \gamma\sigma N(r)\Phi_{ex}(r,t) \quad (24.3)$$

Using a zero-lifetime emission flux $\Phi_{em}^* = L^{-1}\Phi_{em}$, which satisfies the convolution integral as

$$\frac{1}{\tau}\int dt'\Phi^*(r,t')e^{-\frac{t-t'}{\tau}} = \Phi(r,t),$$

we get

$$\left[\frac{1}{c}\frac{\partial}{\partial t} - \nabla \cdot D_{em}(r)\nabla + \mu_a^{em}(r)\right]\Phi_{em}^*(r,t) = \gamma\sigma N(r)\Phi_{ex}(r,t) \quad (24.4)$$

This expression is similar to that of the excitation light in Eq. 24.1. In this case, the light source can be given by the excitation flux Φ_{ex}. The product σN in the right-hand side of Eq. 24.4 shows the absorption coefficient of the absorber (i.e., fluorophore molecules) at the wavelength of excitation $\mu_{a,fluo}^{ex}(r)$. Therefore, this expression gives the amount of energy converted to fluorescence. Conversely, it can give the amount of loss of excitation energy due to the absorption of fluorescence.

From Eqs. 24.1 and 24.4, we get two expressions as follows:

$$\left[\frac{1}{c}\frac{\partial}{\partial t} - \nabla \cdot D_{ex}(r)\nabla + \mu_{a,back}^{ex} + \mu_{a,fluo}^{ex}\right]\Phi_{ex}(r,t) = S_{ex}(r,t) \quad (24.5)$$

$$\left[\frac{1}{c}\frac{\partial}{\partial t}-\nabla\cdot D_{em}(r)\nabla+\mu_{a,\,back}^{em}+\mu_{a,\,fluo}^{em}\right]\frac{1}{\gamma}\Phi_{em}^{*}(r,t)-\mu_{a,\,fluo}^{ex}\Phi_{ex}(r,t)$$

$$=S_{ex}(r,t) \tag{24.6}$$

If the wavelengths of excitation and emission are closed, we can assume that the difference of the diffusion and absorption coefficients can be ignored, namely, the absorption of fluorophore $\mu_{a,back}^{ex}=\mu_{a,back}^{em}$ and $D_{ex}=D_{em}$. Further, the absorption of fluorophore at the emission wavelength can be assumed to be very small. Finally, expressions from Eqs. 24.5 and 24.6 can be simplified as

$$\left[\frac{1}{c}\frac{\partial}{\partial t}-\nabla\cdot D_{ex}(r)\nabla+\mu_{a,\,back}^{ex}(r)\right]\Phi_{total}(r,t)=S_{ex}(r,t) \tag{24.7}$$

where

$$\Phi_{total}=\Phi_{ex}+\frac{1}{\gamma}\Phi_{em}^{*}$$

Equation 24.7 is of the same form as Eq. 24.1, assuming the virtual total energy flux Φ_{total}. The physical meaning of Φ_{total} is the excitation light energy when the fluorophore does not exist.

24.3   Implementation

A laser produces an excitation wavelength of 760 nm and irradiates a fluorescent target at position 0° (excitation point), as shown in Fig. 24.1. The target is 0.5 or 2.0 μM ICG in 1% Intralipid in a cylindrical phantom (L = 60 mm, ϕ = 30 mm). It (4 mm ϕ) is positioned at 11.5 mm from the phantom center. An emission wavelength of 850 nm directs to the IR filter and photodetector. We detect fluorescence at positions 30°, 60°, 90°, 120°, 150°, and 180°.

In Fig. 24.2a, the ICG absorption conversion to fluorescence results in ICG excitation $\Phi_{ex(intralipid + ICG)}$ decay quicker than Intralipid $\Phi_{ex(Intralipid)}$, whereas the lifetime of ICG results in fluorescence $\Phi_{fluo(Intralipid + ICG)}$ broader than excitation $\Phi_{ex(intralipid + ICG)}$.

From theoretical total energy flux Φ_{ex}^{0} (without fluorophores) as

$$\Phi_{ex}^{0}(t)=\Phi_{ex}(t)+\frac{1}{\gamma}\Phi_{fluo}^{*}(t) \tag{24.8}$$

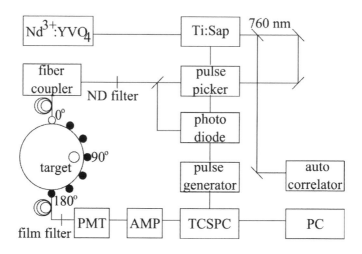

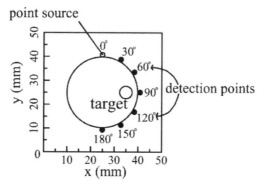

Figure 24.1 Setup of an FDOT system (top) and a sample coordination system (bottom).

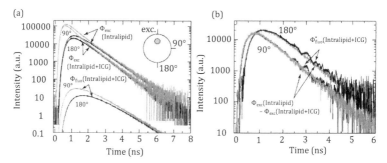

Figure 24.2 Excitation and fluorescence of ICG (left). Zero-lifetime fluorescence and a difference of excitation of ICG (right).

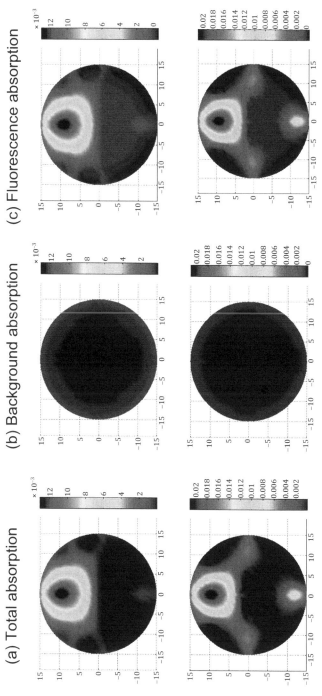

Figure 24.3 Reconstructed images of 0.5 μm ICG (top panel) and 2.0 μM ICG (bottom panel): (a) ICG-Intralipid absorption from excitation, (b) Intralipid absorption from total light, and (c) actual ICG absorption.

in which $\Phi_{ex}(t)$ is the flux of excitation light (with fluorophores), we can calculate the zero-lifetime fluorescence $\Phi^{*}_{fluo}(t)$ as

$$\Phi_{fluo}(t) = \int dt' \Phi^{*}_{fluo}(t')F(t-t') \qquad (24.9)$$

where $\Phi_{fluo}(t)$ is fluorescence and $F(t)$ is the fluorescence decay factor.

In Fig. 24.2b, the excitation difference and the zero-lifetime fluorescence experimentally conform to

$$\Phi^{0}_{ex}(t) - \Phi_{ex}(t) = \alpha\, \Phi^{*}_{fluo}(t) \qquad (24.10)$$

The factor α is the detection efficiency and is calculated by a reconstructed procedure. This is in agreement with the theoretical concept of total light in Eq. 24.8. Therefore, Fig. 24.3 finally shows some results of the reconstructed ICG images.

24.4 Problems

1. Discuss the differences between OCT and FDOT.
2. Explain the concept of total light.
3. Solve the joint partial differential equation of excitation and emission from a diffusion equation.
4. Describe the FDOT system with a block diagram.
5. Give some applications of FDOT.
6. Discuss the weak points of FDOT.

Index